Thomas Peters

Ionization Feedback in Massive Star Formation

Thomas Peters

Ionization Feedback in Massive Star Formation

Simulating the Formation of Massive Stars and Ultracompact H II Regions with Adaptive-Mesh Radiation Hydrodynamics

Südwestdeutscher Verlag für Hochschulschriften

Impressum/Imprint (nur für Deutschland/ only for Germany)
Bibliografische Information der Deutschen Nationalbibliothek: Die Deutsche Nationalbibliothek verzeichnet diese Publikation in der Deutschen Nationalbibliografie; detaillierte bibliografische Daten sind im Internet über http://dnb.d-nb.de abrufbar.

Alle in diesem Buch genannten Marken und Produktnamen unterliegen warenzeichen-, markenoder patentrechtlichem Schutz bzw. sind Warenzeichen oder eingetragene Warenzeichen der jeweiligen Inhaber. Die Wiedergabe von Marken, Produktnamen, Gebrauchsnamen, Handelsnamen, Warenbezeichnungen u.s.w. in diesem Werk berechtigt auch ohne besondere Kennzeichnung nicht zu der Annahme, dass solche Namen im Sinne der Warenzeichen- und Markenschutzgesetzgebung als frei zu betrachten wären und daher von jedermann benutzt werden dürften.

Verlag: Südwestdeutscher Verlag für Hochschulschriften GmbH & Co. KG
Dudweiler Landstr. 99, 66123 Saarbrücken, Deutschland
Telefon +49 681 37 20 271-1, Telefax +49 681 37 20 271-0
Email: info@svh-verlag.de
Zugl.: Heidelberg, U, Diss., 2009

Herstellung in Deutschland:
Schaltungsdienst Lange o.H.G., Berlin
Books on Demand GmbH, Norderstedt
Reha GmbH, Saarbrücken
Amazon Distribution GmbH, Leipzig
ISBN: 978-3-8381-1783-6

Imprint (only for USA, GB)
Bibliographic information published by the Deutsche Nationalbibliothek: The Deutsche Nationalbibliothek lists this publication in the Deutsche Nationalbibliografie; detailed bibliographic data are available in the Internet at http://dnb.d-nb.de.

Any brand names and product names mentioned in this book are subject to trademark, brand or patent protection and are trademarks or registered trademarks of their respective holders. The use of brand names, product names, common names, trade names, product descriptions etc. even without a particular marking in this works is in no way to be construed to mean that such names may be regarded as unrestricted in respect of trademark and brand protection legislation and could thus be used by anyone.

Publisher: Südwestdeutscher Verlag für Hochschulschriften GmbH & Co. KG
Dudweiler Landstr. 99, 66123 Saarbrücken, Germany
Phone +49 681 37 20 271-1, Fax +49 681 37 20 271-0
Email: info@svh-verlag.de

Printed in the U.S.A.
Printed in the U.K. by (see last page)
ISBN: 978-3-8381-1783-6

Copyright © 2010 by the author and Südwestdeutscher Verlag für Hochschulschriften GmbH & Co. KG and licensors
All rights reserved. Saarbrücken 2010

Meinen Eltern
Reimund und Maria Peters
in Dankbarkeit gewidmet

Men and women are not content to comfort themselves
with tales of gods and giants, or to confine their thoughts
to the daily affairs of life; they also build telescopes and
satellites and accelerators, and sit at their desks for
endless hours working out the meaning of the data they
gather. The effort to understand the universe is one of
the very few things that lifts human life a little above the
level of farce, and gives it some of the grace of tragedy.

(Steven Weinberg, The First Three Minutes)

The heavens declare the glory of God;
the skies proclaim the work of his hands.

(Psalms 19, 2)

Contents

List of Figures   vii

List of Tables   ix

List of Abbreviations   xi

List of Symbols   xiii

List of Publications   xvii

1 Introduction   **1**
- 1.1 Basic Concepts of Massive Star Formation 2
  - 1.1.1 Star Formation in a Nutshell 2
  - 1.1.2 The Idea of Bimodality 3
  - 1.1.3 Sites of Massive Star Formation 4
  - 1.1.4 H II Regions around Massive Stars 5
  - 1.1.5 Distribution of Stellar Masses 7
- 1.2 Theoretical Models of Massive Star Formation 8
  - 1.2.1 Monolithic Collapse 8
  - 1.2.2 Competitive Accretion 9
- 1.3 Feedback Effects 10
  - 1.3.1 Small Scale Feedback 10
  - 1.3.2 Large Scale Feedback 12

2 Physics of Star Formation   **15**
- 2.1 Hydrodynamics 15
  - 2.1.1 Euler Equations 15
  - 2.1.2 Thermodynamics of Ideal Gases 17
  - 2.1.3 Sound Waves and Shocks 19
- 2.2 Heating and Cooling 20
- 2.3 Gravity 21
  - 2.3.1 Poisson Equation and Source Terms 21
  - 2.3.2 Gravitational Instability and Jeans Criterion 22
  - 2.3.3 Bonnor-Ebert Spheres 23
  - 2.3.4 Gravitational Collapse of Self-Gravitating Gas 25
- 2.4 Radiation 26
  - 2.4.1 Radiation Fields 26

		2.4.2	Radiative Transfer	28
		2.4.3	Point Sources	31
		2.4.4	Coupling to Hydrodynamics	32
		2.4.5	Ionization	34
		2.4.6	Expansion of H II Regions and Ionization Fronts	37
	2.5	Synthetic Observations		38
		2.5.1	Free-free Radiation	38
		2.5.2	Line Emission	41

3 Simulations of Star Formation   43

	3.1	Overview		43
	3.2	Adaptive Mesh Technique		45
		3.2.1	Data Structure	45
		3.2.2	Refinement Criteria	47
		3.2.3	Prolongation and Restriction	49
	3.3	Operator Splitting		49
	3.4	Hydrodynamics		50
	3.5	Heating and Cooling		52
	3.6	Gravity		53
		3.6.1	External Fields	53
		3.6.2	Self-Gravity	53
	3.7	Radiation		55
		3.7.1	Calculation of Column Densities	55
		3.7.2	Ionizing Radiation	58
		3.7.3	Non-ionizing Radiation	59
	3.8	Sink Particles		60
		3.8.1	Creation and Accretion	60
		3.8.2	Gravitational Field	61
		3.8.3	Equation of Motion	62

4 Results   63

	4.1	Verification		63
	4.2	Driving of Turbulence by Ionization Fronts		65
	4.3	Collapse Simulations		73
		4.3.1	Upper Mass Limit	74
		4.3.2	Star Cluster Formation	76
		4.3.3	Disk Fragmentation	77
		4.3.4	Bipolar Outflows	80
		4.3.5	H II Region Morphologies	88
		4.3.6	Time Variability of UC H II Regions	101

5 Conclusions and Outlook   103

Bibliography   105

List of Figures

2.1	Density profile of a Bonnor-Ebert sphere	24
2.2	Spherically-symmetric isothermal collapse	26
2.3	Zero-age main sequence of the protostellar model	33
2.4	SED of a homogeneous cylinder	41
3.1	Block with guardcells in two dimensions	46
3.2	Adaptive mesh with hierarchy tree	48
3.3	Flow chart of evolution procedure	51
3.4	Multi-grid V-cycle	54
3.5	Long and short characteristics	56
3.6	Calculation of local column densities	57
3.7	Interpolation of column densities	59
4.1	Expansion of D-type ionization front in homogeneous medium	64
4.2	Main stages of clump evaporation	67
4.3	Time evolution of energies and Mach numbers during cloud-crushing	68
4.4	Mach number PDFs during cloud-crushing	69
4.5	Fraction of supersonic gas during cloud-crushing	70
4.6	Slices through the turbulent clump	71
4.7	Line-of-sight mass-weighted histograms of the turbulent clump	72
4.8	Accretion history of single sink simulations	75
4.9	Accretion history of multi sink simulation	78
4.10	Total cluster evolution	79
4.11	Disk fragmentation in multiple sink simulation	81
4.12	Disk fragmentation in single sink simulation	82
4.13	Bipolar outflows in multiple sink simulation	84
4.14	Bipolar outflows in single sink simulation	85
4.15	Ionized accretion flow in observation and simulation	87
4.16	Changes of H II regions	89
4.17	Emitting structures of H II regions	90
4.18	Bubble created by massive protostar	92
4.19	Rotation from face-on to edge-on view	93
4.20	Rotation around polar axis	94
4.21	Different transition angles	95
4.22	Centrally peaked shell for different distances	96
4.23	Different morphologies in Run A	97
4.24	Different morphologies in Run B	98

List of Figures

4.25 Abnormal SEDs in Run B 100
4.26 Time variability of UC H II regions 102

List of Tables

1.1	Classification of H II region sizes	5
1.2	Morphological types of UC H II regions	6
2.1	VLA telescope parameters	40
4.1	Overview of collapse simulations	74
4.2	UC H II region morphologies in surveys and simulations	99

List of Abbreviations

AMR	adaptive mesh refinement, page 43
CMF	clump mass function, page 7
FWHM	full width at half maximum, page 40
H I	neutral hydrogen, page 3
H II	ionized hydrogen, page 3
IMF	initial mass function, page 7
ISM	interstellar medium, page 1
MHD	magnetohydrodynamics, page 11
PDF	probability density function, page 66
SED	spectral energy distribution, page 40
SFE	star formation efficiency, page 7
SFR	star formation rate, page 7
SPH	smoothed particle hydrodynamics, page 43
UC H II region	ultracompact H II region, page 5
VLA	Very Large Array, page 40
ZAMS	zero-age main sequence, page 32

List of Symbols

A_c	collisional ionization rate, page 34
α_ν	absorption coefficient, page 28
α_R	radiative recombination rate, page 34
A_p	photoionization rate, page 34
β	ratio of rotational to gravitational energy, page 25
B_ν	specific intensity of black body, page 30
c	speed of light, page 27
c_P	specific heat at constant pressure, page 18
c_s	sound speed, page 19
c_V	specific heat at constant volume, page 17
E	radiation energy, page 26
e	specific internal energy, page 16
ϵ	internal energy density, page 16
ϵ_{tot}	total energy density, page 16
e_{tot}	total specific energy, page 16
F	total flux, page 27
F_ν	specific flux, page 26
G	Newton's constant, page 21
$\boldsymbol{g}$	gravitational acceleration, page 21
γ	adiabatic index, page 18
Γ_{acc}	accretion heating rate, page 34
Γ_d	dust heating rate, page 33
Γ_{ph}	photoionization heating rate, page 32
Γ_{st}	stellar heating rate, page 33
h	Planck's constant, page 27
h	specific enthalpy, page 18
I	total intensity, page 27
I_ν	specific intensity, page 26

List of Symbols

J	total mean intensity, page 27	
J_ν	specific mean intensity, page 27	
j_ν	emission coefficient, page 28	
κ_ν	opacity coefficient, page 29	
κ_P	Planck mean opacity, page 30	
k_B	Boltzmann's constant, page 17	
k_J	Jeans wave number, page 22	
L	stellar luminosity, page 32	
L_{acc}	accretion luminosity, page 34	
Λ_{dust}	dust cooling rate, page 20	
Λ_{gd}	gas-dust coupling, page 20	
λ_J	Jeans length, page 22	
Λ_{mol}	molecular line cooling rate, page 20	
$\mathcal{M}$	Mach number, page 19	
$\dot{M}$	accretion rate, page 34	
M_J	Jeans mass, page 22	
M_{max}	maximum mass of star in cluster, page 9	
m_p	proton mass, page 17	
$M_\odot$	the solar mass, page 3	
M_{tot}	total mass of stars in cluster, page 9	
μ	mean molecular weight, page 17	
N	column density, page 29	
$\mathbf{n}$	direction vector, page 26	
n_e	electron number density, page 34	
n_H	hydrogen number density, page 30	
n_{HI}	neutral hydrogen number density, page 30	
n_{HII}	ionized hydrogen number density, page 30	
ν	frequency, page 26	
ν_T	hydrogen ionization threshold frequency, page 29	
Ω	solid angle, page 26	
P	gas pressure, page 16	
ϕ	gravitational potential, page 21	
ψ	photon number density, page 27	
R	ideal gas constant, page 17	
r_{acc}	accretion radius, page 34	
ρ	mass density, page 16	

List of Symbols

ρ_{max}	sink particle threshold density, page 60
R_S	Strömgren radius, page 37
r_{sink}	sink particle radius, page 60
$R_\odot$	the solar radius, page 31
s	path length, page 28
σ	Stefan-Boltzmann constant, page 30
σ_ν	absorption cross section, page 28
T	gas temperature, page 17
t	time, page 16
T_{acc}	temperature of accretion radiation, page 34
τ_ν	optical depth, page 29
t_{ff}	free-fall time, page 25
T_{star}	temperature of star, page 32
u	total radiation energy density, page 27
u_ν	specific radiation energy density, page 27
v	gas velocity, page 16
x_{HI}	fraction of neutral hydrogen, page 30
x_{HII}	fraction of ionized hydrogen, page 30

List of Publications

The results presented in this thesis have been published in the following papers:

1. T. Peters, R. Banerjee and R. S. Klessen. "Ionization front-driven turbulence in the clumpy interstellar medium". *Physica Scripta*, vol. T132(014026), 2008.

2. T. Peters, R. Banerjee, R. S. Klessen, M.-M. Mac Low, R. Galván-Madrid and E. R. Keto. "H II regions: Witnesses to Massive Star Formation". *The Astrophysical Journal*, vol. 711, 1017–1028, 2010.

3. T. Peters, M.-M. Mac Low, R. Banerjee, R. S. Klessen and C. P. Dullemond. "Understanding Spatial and Spectral Morphologies of Ultracompact H II Regions". *The Astrophysical Journal*, vol. 719, 831–843, 2010.

4. T. Peters, R. S. Klessen, M.-M. Mac Low and R. Banerjee. "Limiting Accretion onto Massive Stars by Fragmentation-Induced Starvation". *The Astrophysical Journal*, vol. 725, 134–145, 2010.

1 Introduction

> Das Schönste, was wir erleben können, ist das Geheimnisvolle. Es ist das Grundgefühl, das an der Wiege von wahrer Kunst und Wissenschaft steht. Wer es nicht kennt und sich nicht mehr wundern, nicht mehr staunen kann, der ist sozusagen tot und sein Auge erloschen.
>
> *(Albert Einstein, Mein Weltbild)*

Massive stars have a strong impact on the Universe. With masses exceeding 100 solar masses and lifetimes of only a few million years, they produce all the heavy elements during their lifetimes. They end their lives in powerful supernova explosions, which release enormous amounts of energy and momentum and inject enriched material back into the ambient interstellar medium (ISM), where the next generation of stars is already waiting to be born. These supernova explosions also supplied the material from which our own Solar System, the Earth and all living creatures formed.

Despite the utter importance of massive stars for the matter and energy cycle in the Galaxy, very little is known about their origin. We know that the formation of massive protostars requires high accretion rates. They begin hydrogen burning while they are still in the main growth phase. This leads to very high luminosities in the optical and ultraviolet. It turns out that this radiation feedback is a crucial point in understanding massive star formation. In particular, it is an open question whether radiation feedback is a limiting mechanism that determines how much mass nascent stars can attain.

Once the protostar is luminous enough to emit a considerable fraction of high energy photons, it begins to ionize the gas in its neigborhood. An H II region[1] forms. Observations of H II regions around massive stars pose some puzzling questions. H II regions show a variety of different morphologies, which are difficult to understand theoretically. A statistical analysis shows that there far too many very small (ultracompact, UC) H II regions found compared to the predictions from simple models of expanding ionized bubbles. Finally, some H II regions even shrink or change their morphology within decades. Other interesting questions concern the relation between H II regions and bipolar outflows emerging from massive protostars.

In this work, we address these questions with numerical simulations of massive star formation. We present the first three-dimensional collapse simulations that include the radiative feedback by both ionizing and non-ionizing radiation. The results lead to a new conceptual understanding of the formation of massive stars and a consistent interpretation of the observational findings.

---

[1] H II denotes ionized hydrogen, H I neutral hydrogen.

1 Introduction

In this chapter, we provide some background on star formation (section 1.1), theoretical models for massive star formation (section 1.2) and feedback processes (section 1.3). Chapter 2 presents the underlying fundamental physics. The simulation method is explained in chapter 3. Chapter 4 shows the results of the simulations. Finally, in chapter 5 we summarize the conclusions and give an outlook on possible future directions of research.

1.1 Basic Concepts of Massive Star Formation

Before we delve into the intricacies associated with massive star formation, we summarize some general facts and concepts which are useful when studying massive star formation. After a quick review of low-mass star formation in section 1.1.1, we investigate the differences between low-mass and high-mass star formation in section 1.1.2. We then discuss some observational aspects (section 1.1.3) with a focus on H II regions (section 1.1.4) and the stellar mass spectrum (section 1.1.5).

1.1.1 Star Formation in a Nutshell

In this section we give a brief overview of the standard picture of low-mass star formation. This will enable us in the next section to assess whether high-mass star formation should be expected to proceed along the same lines or whether we might anticipate some differences. Since this work is only concerned with star formation at the scale of individual molecular clouds[2] and downwards, we will only review the collapse of single molecular cloud cores and not examine star formation at the galactic scale.

The now-classical theory of low-mass star formation was set out by Shu et al. (1987). A more up-to-date summary with a different appreciation of the involved physics is given by Mac Low & Klessen (2004), but the basic phases have not changed much. The initial phase is a condensation of gas in the interstellar medium that becomes gravitationally unstable. This is typically due to compressive turbulent motions in the ISM. This bound core will then collapse isothermally until it loses its ability to cool efficiently. This happens at number densities of $n_{H_2} \approx 10^{10}\,\mathrm{cm}^{-3}$, where the gas becomes optically thick and the excess energy due to the compression cannot be radiated away anymore. Then the gas heats up until a temperature of $T \approx 2000\,\mathrm{K}$ is reached, which corresponds to the dissociation energy of H_2 molecules. Dissociation of H_2 absorbs the excess energy, such that collapse can continue until all H_2 is dissociated. The prestellar phase comes to an end, and a protostar has formed.

The protostellar evolution proceeds in subsequent phases which can be distinguished observationally by their spectral energy distribution (SED). The SED and in particular its infrared excess shows different forms and slopes as function of wavelength depending on the protostellar class of the object (André et al. 2000, Stahler & Palla 2004). In the first phase (class 0), the protostar still grows in mass by accreting gas from its

---
[2]Giant molecular clouds can reach several 10 pc in size (1 pc $\approx 3.1 \cdot 10^{18}$ cm).

surrounding envelope through an accretion disk[3]. The class I phase starts when the protostar begins to drive outflows which create cavities in the envelope. The envelope is totally removed in the class II phase and the protostar has stopped accretion. The protostar now starts its pre-main-sequence evolution. The main source of energy for pre-main-sequence stars is gravitational contraction. When the central temperature is sufficient to ignite hydrogen fusion, a main sequence star is born.

1.1.2 The Idea of Bimodality

The preceding overview reflects the formation of stars with mass[4] $M \lesssim 7 M_\odot$. We call them low-mass stars. For stars more massive than this, two relevant timescales become so short that the outlined scenario may change: the Kelvin-Helmholtz timescale of the pre-main-sequence evolution and the main sequence lifetime.

As long as nuclear reactions have not yet been triggered, the protostar compensates for its energy losses by gravitational contraction. The timescale for this Kelvin-Helmholtz phase is

$$t_{\text{KH}} = \frac{GM^2}{RL} \tag{1.1}$$

with Newton's constant G, the protostellar mass M, the protostellar radius R and the luminosity L (Huang & Yu 1998). Generally, the luminosity scales with mass to a power larger than 2, so that t_{KH} becomes smaller as M increases. The exact evolutionary times can only be found by numerically solving stellar evolution models. Iben (1965) found that the pre-main-sequence lifetime for a $1 M_\odot$ star is $5 \cdot 10^7$ yr, while a $15 M_\odot$ star only needs $6 \cdot 10^4$ yr to start hydrogen burning. This means that, unless the $15 M_\odot$ star forms with an accretion rate exceeding $2.5 \cdot 10^{-4} M_\odot \, \text{yr}^{-1}$, the star begins nuclear fusion while it is still accreting more gas (Keto & Wood 2006). Generally, stars with $M \gtrsim 8 M_\odot$ continue accretion while they are already burning hydrogen (Palla & Stahler 1993, Yorke & Bodenheimer 2008).

Not only the pre-main-sequence evolution of massive stars is quicker than for low-mass stars, also the time on the hydrogen main sequence becomes much shorter. The evolutionary time on the main sequence scales with the stellar mass like (Huang & Yu 1998)

$$t_{\text{H}} \propto M^{-2.2}. \tag{1.2}$$

Thus, in order to stay on the main sequence long enough, a massive star must be continuously supplied with material from its surroundings. Otherwise, it will very quickly leave the main sequence, and the ejection of its envelope will stop the accretion process (Keto & Wood 2006).

The briefness of these two timescales shows that massive stars must accrete further material while they are already burning hydrogen. This is the distinctive feature of massive star formation. The nuclear fusion gives rise the several radiative feedback processes of the star on its accretion flow. It is the interaction between the accretion

---

[3]Circumstellar disks typically have radii up to $1000 \, \text{AU}$ ($1 \, \text{AU} \approx 1.5 \cdot 10^{13} \, \text{cm}$).
[4]The solar mass is $M_\odot = 1.989 \cdot 10^{33} \, \text{g}$.

flow and the radiation field of the massive protostar that makes high-mass star formation different from low-mass star formation and needs to be understood. In particular, ionizing radiation creates regions of ionized gas, H II regions, around the massive star. These regions are well studied (see section 1.1.4) and can be used to test the numerical simulations. The existence of an H II region is the most prominent feature of the bimodality of low-mass and high-mass star formation.

1.1.3 Sites of Massive Star Formation

The high accretion rates required to form massive stars suggest that special initial conditions might be necessary. Indeed, the idea of bimodal star formation originated in work by Herbig (1962), who found discrepancies between Taurus with no stars larger than $2M_\odot$ and Orion with both low-mass and high-mass stars. Surveys of the Galactic disk (Solomon et al. 1985) find that massive stars can only form in the most massive clouds, which are typically located in Galactic spiral arms, whereas low-mass star formation is distributed more uniformly throughout the Galaxy.

Of course, this bimodality in the distribution of the sites of massive star formation does not yet answer the question whether massive stars require a different formation mechanism than low-mass stars to overcome the radiation feedback. Unfortunately, direct observations of massive star accretion are hindered by the dense envelopes that surround massive protostars and make it hard to look at the accretion process in detail. Additionally, massive stars are rare and live only for a short time, so that the important stages of massive star formation are statistically hard to find, and the sites of massive protostars are far away (Evans 1999).

Turbulent motions in a molecular cloud will typically trigger fragmentation of the cloud into several gravitationally bound cores. If these cores are gravitationally unstable, they will collapse and form stars. Observations show that the cores associated with high-mass star formation are denser and more massive (Motte et al. 2008) than their low-mass counterparts (Myers et al. 1986). These high densities may be necessary to obtain sufficiently high accretion rates for massive star formation. On the other hand, a massive accretion flow is susceptible to gravitational fragmentation, which in turn may lead to the formation of multiple stars (Klessen & Burkert 2000, Kratter & Matzner 2006). This could explain the fact that massive stars are only rarely found in isolation (Ho & Haschick 1981) and that they have a higher number of companion stars compared to low-mass stars (Zinnecker & Yorke 2007). Of course, these additional stars can significantly change the dynamics of the accretion flow.

Chini et al. (2004) and Nürnberger et al. (2007) report observations of a $20M_\odot$ protostar embedded in a $100M_\odot$ accretion disk. The presence of a jet indicates that the protostar is still accreting gas, and the accretion rate is estimated to be $\dot{M} \approx 10^{-4} M_\odot \, \mathrm{yr}^{-1}$. This is very high compared to accretion rates of $\dot{M} \approx 10^{-7}$ to $10^{-6} M_\odot \, \mathrm{yr}^{-1}$ as found in low-mass star formation but suggests that star formation at protostellar masses around $20M_\odot$ is still similar to low-mass star formation and that feedback effects do not yet hinder accretion through the disk.

Once an H II region forms around a protostar, there is a competition between the

Class of Region	Diameter (pc)	Density (cm^{-3})	Emission Measure (pc cm^{-6})	Ionized Mass ($M_\odot$)
Hypercompact	$\lesssim 0.03$	$\gtrsim 10^6$	$\gtrsim 10^{10}$	$\sim 10^{-3}$
Ultracompact	$\lesssim 0.1$	$\gtrsim 10^4$	$\gtrsim 10^7$	$\sim 10^{-2}$
Compact	$\lesssim 0.5$	$\gtrsim 5 \cdot 10^3$	$\gtrsim 10^7$	~ 1
Classical	~ 10	~ 100	$\sim 10^2$	$\sim 10^5$
Giant	~ 100	~ 30	$\sim 5 \cdot 10^5$	$10^3 - 10^6$
Supergiant	> 100	~ 10	$\sim 10^5$	$10^6 - 10^8$

Table 1.1: Classification of H II region sizes, adopted from Kurtz (2005). The table shows the diameters D, the electron number densities n_e, the emission measures EM and typical ionized gas masses within the regions. Massive stars form strongly clustered, so that even UC H II regions can be powered by more than one massive star.

expansion of the ionized material and the accretion of gas onto the protostar. Keto (2002a) observed an H II region around a massive protostar which shows evidence of accreting ionized gas. Keto & Wood (2006) further analyzed this region and concluded that the ionization feedback will not cut off accretion. These observations represent a weaker version of an idea put forward by Walmsley (1995), who suggested that an H II region can be quenched if accretion is strong enough. Observations by Beltrán et al. (2006) point in a similar direction. They also found infall onto a massive protostar although it is surrounded by an H II region.

1.1.4 H II Regions around Massive Stars

H II regions are found in various shapes and sizes. Table 1.1 shows the common classification of sizes of H II regions. The selection criteria are the diameter D of the region, the electron number density n_e and the emission measure

$$\mathrm{EM} = \int_0^D n_e^2 \, \mathrm{d}r. \tag{1.3}$$

Smaller regions are generally denser, but contain less ionized gas than larger regions. It seems natural to assume that the classification represents an evolutionary sequence of the expanding H II region: When the massive star forms, a hypercompact H II region develops, which then expands to become an ultracompact (UC) H II region. Since massive stars form in clusters, expanding UC H II regions typically merge with other UC H II regions to form compact or even larger H II regions. Investigations of UC H II regions are an important tool in understanding massive star formation (Churchwell 2002).

UC H II regions show different types of morphologies (for the definitions see table 1.2). A classification of these morphologies was first introduced by Wood &

1 Introduction

Type	Characteristics	WC89	K94	D05
Spherical	symmetric, centrally peaked	24%	36%	21%
Cometary	bright edge and extended tail	20%	16%	14%
Core-halo	compact peak with fainter halo	16%	9%	—
Shell-like	(possibly broken up) ring of emission	4%	1%	28%
Irregular	several peaks with common halo	17%	19%	11%
Bipolar	elongated	—	—	8%
Unresolved	emission peak of beam size	19%	19%	18%

Table 1.2: Morphological types of UC H II regions. The original morphological classification is due to Wood & Churchwell (1989) and has later been extended by the bipolar type by De Pree et al. (2005). Surveys of several massive star forming regions (Wood & Churchwell 1989 (WC89), Kurtz et al. 1994 (K94), De Pree et al. 2005 (D05)) show different relative frequencies of the individual types.

Churchwell (1989), who studied 75 UC H II regions and found that they fall into the groups spherical or unresolved (43%), cometary (20%), core-halo (16%), shell (4%) and irregular (17%). Table 1.2 shows that similar studies by Kurtz et al. (1994) and De Pree et al. (2005) led to totally different relative frequencies. In addition, De Pree et al. (2005) abandoned the core-halo category and introduced the new bipolar one. Higher resolution and sensitivity images had shown that most H II regions are surrounded by halos, which rendered the core-halo morphology redundant. Instead, better observations had revealed that many UC H II regions have an elongated form (Churchwell 2002). The morphological classification can also be applied to compact H II regions (Garay et al. 1993).

The existence of different morphological types gives rise to the question of their physical origin. In particular, the different relative frequencies in the surveys may be related to different physical conditions in these regions. To explain the curious shape of cometary H II regions, it was suggested that they might be bow shocks created by stellar winds of stars moving supersonically through their natal molecular clouds (Van Buren et al. 1990, Mac Low et al. 1991, Van Buren & Mac Low 1992). However, the high stellar velocities required by bow shock models are problematic (Mac Low et al. 2007).

Wood & Churchwell (1989) first noticed what is now called the UC H II lifetime problem: They compared an analytical estimate, based on the sound speed of the ionized gas, of the expansion time of an H II region around a massive protostar with the total main-sequence lifetime of an O star and found that the amount of UC H II regions in their statistical sample of O stars was incomprehensibly large. They concluded that the H II region must be trapped by some unknown mechanism and proposed that the expansion may be inhibited by infalling gas.

Since its discovery, a number of suggestions have been made to resolve the UC H II lifetime problem (Mac Low 2008). The original idea of ram pressure confinement by infalling gas by Wood & Churchwell (1989) suffers from instabilities in this setup. The density gradients present in this situation do not allow a steady state solution (Yorke 1986, Hollenbach et al. 1994), and the boundary between the ionized and the neutral gas will be subject to Rayleigh-Taylor instabilites (Mac Low 2008). De Pree et al. (1995) put forward the idea that the H II regions may be confined by external thermal pressure. This idea was developed in more detail by García-Segura & Franco (1996). However, sufficiently high pressures require densities that would lead to gravitational collapse within a free-fall time, whereas the H II region would have to be confined for more than three free-fall times (Mac Low 2008). Xie et al. (1996) claimed that the demanded thermal pressures would lead to emission measures higher than supported by observations. They instead argued that confinement by turbulent pressure should be considered. But turbulence under these conditions would decay quickly (Stone et al. 1998, Mac Low 1999) and is difficult to replenish for several free-fall times (Mac Low 2008).

Another curious property of UC H II regions is their time variability. Repeated observations of the same H II regions over several decades (Franco-Hernández & Rodríguez 2004, Rodríguez et al. 2007, Galván-Madrid et al. 2008, Gómez et al. 2008) show variations in the size and flux of the H II regions by 2–9 % per year and even morphological changes. These observations indicate that H II regions are highly dynamical objects and can be interpreted to display ionized accretion flows.

1.1.5 Distribution of Stellar Masses

Stars with $M \gtrsim 8 M_\odot$ begin hydrogen burning during the accretion phase. The distribution of stellar masses in the present-day universe has its upper end more than 20 times above this value. For example, Barniske et al. (2008) infer that the initial masses of two of the most luminous stars in the Milky Way, WR 102ka and WR 102c, are $150 \lesssim M/M_\odot \lesssim 200$ and $100 \lesssim M/M_\odot \lesssim 150$, respectively.

The statistical distribution of stellar masses at birth within a stellar cluster is called the initial mass function (IMF) $N(M)$. Although there is much discussion about its origin, universality and best parametrization (Bonnell et al. 2007, Pudritz 2002, Kroupa 2002, Massey 1998, Chabrier 2003), there is a general agreement on the high-mass tail of the IMF, which is given by the famous Salpeter slope (Salpeter 1955) of $N(M) \propto M^{-2.35}$. Given that slope, Figer (2005) analyzed the Arches star cluster, which is sufficiently large that very massive stars should be found and sufficiently young that these stars have not yet been destroyed by supernova explosions. However, he did not find any star more massive than $130 M_\odot$, although statistically 18 were expected. With the help of Monte Carlo simulations he tried to reproduce the observed IMF in the cluster and suggested a high-mass cut-off of the IMF at $150 M_\odot$.

Interestingly, there seems to be a similarity between the distribution of stellar masses and molecular cloud cores. Molecular clouds are highly substructured objects with larger-scale subunits (clumps) and smaller condensations (cores). In a detailed study of the Pipe Nebula, Alves et al. (2007) found that its core mass function (CMF) has a shape very similar to the IMF, just scaled up to higher masses by a factor of about

3. This suggests a direct relation between CMF and IMF: A fraction of roughly 30 % of the gas available in the cores is converted into stars. This fraction of gas is called the star formation efficiency (SFE, not to be confused with the star formation rate (SFR), the mass converted into stars per unit time). However, the CMF statistics only contained cores between 0.5 and 28 $M_\odot$. Whether higher mass cores lead to the formation of high-mass stars or to which degree they fragment is a subject of ongoing debate (Krumholz & Bonnell 2007, Clark et al. 2008). In any case, the similarity between CMF and IMF, wich is also found in other studies (Motte et al. 1998, 2001, Beuther & Schilke 2004) should be explained by theories of star formation.

1.2 Theoretical Models of Massive Star Formation

There are two main schools of thought in the theory of massive star formation (Zinnecker & Yorke 2007, Krumholz & Bonnell 2007). They differ primarily in the way the gas that forms the massive star is assembled and say little about how exactly the massive star forms. In some sense, they are different theories of fragmentation of the original molecular cloud. Nevertheless, the question of the initial conditions for massive star formation and the fragmentation of the parental molecular cloud is an important part of the full problem.

The first point of view is called monolithic collapse or core accretion and is summarized in section 1.2.1. Here, the mass that can be converted into stars is assembled before star formation sets in. The alternative competitive accretion model is presented in section 1.2.2. In this scenario, mass is gathered during the whole duration of the star formation process.

1.2.1 Monolithic Collapse

The core accretion model is strongly inspired by the relation between CMF and IMF presented in section 1.1.5. The idea is that collapsing cores lead to the formation of either single stars or bound stellar systems like binaries. The more massive the core, the more massive the star (McKee & Tan 2002, 2003). However, this poses the problem that a single object must form from a core which may have several hundred Jeans masses and thus would naturally fragment into hundreds of individual objects of roughly one Jeans mass (Krumholz & Bonnell 2007). Dobbs et al. (2005) performed numerical simulations assuming isothermal and non-isothermal collapse of a turbulent molecular cloud core with several Jeans masses and found substantial fragmentation as expected.

A possible solution to this problem is the radiative feedback by the first stars that form in the core (Krumholz 2006, Krumholz et al. 2007a). Their stellar and accretion luminosities heat the gas up and thereby shift the Jeans mass towards higher masses. Indeed, the gas temperatures of Krumholz et al. (2007a) are generally higher than the temperatures in the non-isothermal calculations of Dobbs et al. (2005) and therefore suppress fragmentation much stronger. With the help of this mechanism, Krumholz et al. (2009) could form a 43 $M_\odot$ star from a 100 $M_\odot$ core. However, it is still an open question to which degree the results depend on the specific initial conditions chosen in these simulations. The core of Krumholz et al. (2009) is centrally concentrated and set up in solid-body rotation, but there is no turbulent velocity field superimposed on the rotation. These conditions of course prefer the formation of a small number of objects.

1.2.2 Competitive Accretion

In contrast to the core accretion model, which approaches the problem at the scale of cores, the competitive accretion model starts with a turbulent molecular cloud and models the formation of a full cluster of stars, including massive ones. Indeed, simulations of turbulent molecular clouds show strong fragmentation and can easily reproduce the IMF for low-mass stars (Klessen et al. 1998, Bate et al. 2003). The low-mass stars formed during the initial fragmentation phase then form small multiple systems and competitively accrete from a common gas reservoir (Bonnell et al. 2001). It is the combined gravitational potential of the star cluster that determines how massive the stars can become (Bonnell et al. 2007). Typically, the deepest potential well is located in the center of the cluster, so that the gas is funneled towards the stars located there which then become the most massive (Bonnell et al. 1997, 2001). Observations of embedded clusters (Lada & Lada 2003) show that in many cases the most massive stars are in fact located at their center, altougth there are exceptions. The similarity between the CMF and the IMF is explained as a natural outcome of the fragmentation of the turbulent cloud, without any one-to-one correspondence between cores and stars (Krumholz & Bonnell 2007). This is supported by the analysis of Lada et al. (2008), which shows that most cores in the Pipe Nebula are not gravitationally bound and hence will not collapse to form stars. It is also possible that stars from the multiple systems which form in the course of the fragmentation merge to form massive stars, which would circumvent the problem of radiation feedback (Bonnell et al. 1998, Stahler et al. 2000).

In models of competitive accretion, the evolution of the stellar cluster is intimately connected with the evolution of the most massive stars. Bonnell et al. (2004) found a correlation between the mass of the most massive object in the star cluster, M_{max}, and the total cluster mass, M_{tot}. These two masses roughly followed the scaling

$$M_{\mathrm{max}} \propto M_{\mathrm{tot}}^{2/3} \qquad (1.4)$$

over the simulation runtime. This relation is a direct prediction of the competitive accretion model that can be compared with observations. However, due to the uncertainties in the observed data, this power law is difficult to distinguish from other proposed functional forms of the relation between M_{max} and M_{tot} (Weidner et al. 2009).

1.3 Feedback Effects

Radiative feedback indispensably accompanies massive star formation. The problem can be divided into the effects of radiation pressure from non-ionizing radiation on dust grains and the effects of ionizing radiation on the molecular gas. Both classes of feedback have been studied in the context of massive star formation (Mac Low 2008, Zinnecker & Yorke 2007, Klessen et al. 2009). We begin in section 1.3.1 with radiative feedback on small scales as it is relevant for massive star formation. We continue in section 1.3.2 with feedback on larger scales, with a focus on the effects of ionizing radiation on the parental molecular cloud.

1.3.1 Small Scale Feedback

Any viable theory of massive star formation must explain how the massive protostar can manage to continue accretion although it exerts a high radiation pressure on the inflowing material. That radiation pressure poses a problem was first realized by Kahn (1974). He performed semi-analytical calculations with spherical symmetry including radiation-dust interaction. Dust has a high opacity, but if it comes sufficiently close to the protostar that it melts, the opacity drops and radiation does not couple any more to the infalling material. Otherwise, a cocoon forms which radiation cannot penetrate, and accretion onto the protostar stops. Kahn argued that the mass inflow rate $\dot{M}$ must lie between two critical values: If $\dot{M}$ is too low, then the ram pressure of the inflowing material is not high enough to reach the vicinity of the star where dust melts; if $\dot{M}$ is too high, then the material is so opaque that the radiation pressure poses an unsurmountable barrier. He estimated these critical values to coincide, which led him to a stellar mass limit of $40M_\odot$. Wolfire & Cassinelli (1987) basically confirmed this picture with spherically symmetric numerical calculations. They studied different scenarios with varying dust grain properties and found a strong dependence on the grain destruction temperature and grain sizes. Yorke & Krügel (1977) also performed spherically symmetric collapse simulations with radiation feedback where accretion onto the massive protostar could be stopped. Edgar & Clarke (2004) found that radiation pressure can also stop Bondi-Hoyle accretion already at $10M_\odot$.

Jijina & Adams (1996) considered the effect of radiation pressure as an effective potential in an infall model put forward by Ulrich (1976). This model includes non-spherical accretion, and in their calculations they find certain conditions under which they can even accrete $100M_\odot$. Nakano (1989) found that non-spherical accretion onto a $100M_\odot$ star can overcome the radiation pressure at accretion rates 50 times smaller than in the spherical case.

The problem is further mitigated in two-dimensional simulations with cylindrical symmetry. Yorke & Bodenheimer (1999) found that the radiation flux can differ by orders of magnitude between pole-on and edge-on views of the accretion disk, the so-called flashlight effect. Hence, if the radiation can escape perpendicular to the disk, the radiation pressure in the equatorial plane decreases. Yorke & Sonnhalter (2002) showed that the flashlight effect is stronlgy enhanced by frequency-dependent

1.3 Feedback Effects

treatment of the radiation transfer problem. This is because the short wavelength components of the radiation field, which are the most important for the acceleration of dust grains, are more concentrated towards the poles than the longer wavelength components, which are nearly isotropically distributed. For example, the collapse of a $120M_\odot$ clump led to a $43M_\odot$ star with frequency-dependent radiation transfer but only to a $23M_\odot$ star if radiation transport was gray.

The flashlight effect is amplified if outflows are present. Krumholz et al. (2005b) performed static three-dimensional radiative transfer simulations. They set up a $50M_\odot$ star with a $5M_\odot$ accretion disk inside a $50M_\odot$ envelope and analytically prescribed an outflow cavity. Depending on the opening angle and the curvature of the cavity, the radiation pressure was reduced at least by a factor of four. Banerjee & Pudritz (2007) found such outflow cavities in self-consistent magneto-hydrodynamical (MHD) simulations of collapsing dense cores.

In fully three-dimensional radiation-hydrodynamical calculations, additional effects set in. Krumholz et al. (2005a) simulated the collapse of rotating dense cores and found that radiation pressure begins to build up asymmetric bubbles of gas when the protostar reaches $17M_\odot$. The asymmetry is the result of a self-amplifying effect: If the bubble is more elongated in one direction than in the others, the optical depth in this direction is smaller. Hence, radiation is collimated towards this direction, enlargening the elongation by concentrated radiation pressure even more. Although inflowing gas cannot cross the bubble, accretion never stops. This is because the gas finds its way along the edge of the bubble onto the protostar. At some point, the boundary between the dense envelope and the underdense interior of the bubble becomes Rayleigh-Taylor unstable. Then the bubble gets destroyed, and the gas flows unhindered towards the protostar. A caveat to this scenario is that Krumholz et al. (2005a) used a gray flux-limited diffusion scheme to solve the radiative transfer problem. If the bubbles persist in a multi-frequency treatment, when the optical depth is not the same for all photons and hence the radiation pressure acts over a larger volume of gas, is an open question.

Quite intricate effects of radiation-hydrodynamics appear in the presence of magnetic fields. Arons (1992) used linear stability theory to show that a magnetized atmosphere of a neutron star can become overstable[5]. Gammie (1998) extended the linear stability analysis to radiation-dominated accretion disks. This photon bubble instability has been found numerically in simulations of accretion disks around black holes (Turner et al. 2005) and massive protostars (Turner et al. 2007). The mechanism that drives the instability transfers energy from the radiation field to the magnetic medium. The starting point are compressive MHD waves, which generate density fluctuations. The radiation tends to escape through the regions of lower density, which leads to an acceleration of the gas along the magnetic field lines and in turn evacuates the low-density regions. In this way, the amplitude of the wave grows with time. A train of propagating shocks is generated, and the radiation can escape through the regions of low density between these shocks.

All the effects mentioned above are produced in some way by the radiation pressure

---

[5] An overstability is an instability that results in oscillations of growing amplitude.

of non-ionizing radiation. However, the feedback by ionizing radiation may as well be important, although so far there are only few numerical studies (Krumholz & Bonnell 2007, Mac Low 2008). Larson & Starrfield (1971) showed by analytical estimates that ionization feedback could indeed be the most important feedback effect. This is supported by calculations by Nakano et al. (1995), who found that ionizing radiation can stop accretion onto a $100\,M_\odot$ star. Generally, ionization heats the interstellar gas up to $10^4\,\mathrm{K}$, which corresponds to sound speeds around $10\,\mathrm{km\,s^{-1}}$. If this speed is larger than the escape speed from the protostar, then the H II region may expand and cut off accretion. Semi-analytical models of ionization feedback around a massive protostar (Keto 2002b, 2007) show some basic features like expanding H II regions and effects on the accretion disk, but they neglect the highly dynamical and asymmetric interaction between the approaching gas and the ionization shock front. Direct numerical simulations are inevitable to realistically model ionization feedback. One may also speculate whether, in analogy to non-ionizing radiation, "ionization-driven Rayleigh-Taylor instabilities" or "ionization bubbles" can be found.

1.3.2 Large Scale Feedback

While radiation pressure is the feedback process that has been studied the most in massive star formation, it can generally be neglected on larger scales. Krumholz & Matzner (2009) found that, unless one is interested in the Galactic center or starburst galaxies, the thermal pressure is dominant in H II regions.

On molecular cloud scales, ionization feedback by massive stars dramatically changes the places in which they have formed. Whitworth (1979) estimated that expanding H II regions around massive stars could erode large parts of their parental molecular cloud. An SFE of 4 % for a massive cloud of $10^5\,M_\odot$ would suffice to completely disperse the remaining 96 % of the cloud.

This scenario is a typical example for negative feedback: The SFE with feedback is smaller than the (hypothetical) SFE without feedback. One important question is whether feedback can also be positive. This would mean that the feedback triggers star formation that would otherwise not have occured.

Elmegreen & Lada (1977) put forward the idea that expanding ionization fronts might sweep up interstellar gas until star formation is possible in the dense cold shell surrounding the ionization front. Numerical simulations of dynamically expanding H II regions by Dale et al. (2007a) indeed find that the surrounding shell fragments and forms gravitationally bound objects. The properties of these fragments agree well with analytical work on this "collect and collapse model" by Whitworth et al. (1994).

However, massive stars do not form alone, and not in a quiescent medium. Much more realistic simulations by Dale et al. (2005) of ionization feedback in star cluster formation show evidence for both positive and negative feedback. However, these simulations do not have enough resolution to follow the star formation process. Dale et al. (2007b) studied the effects of ionization feedback from an external source on a turbulent molecular cloud. They again found positive and negative feedback, but the overall balance showed an increase of 1 % of the SFE in comparison to a control run without feedback. They also found no way to observationally differentiate between triggered and revealed[6] star formation. Mac Low et al. (2007) found as well that expanding H II regions in a turbulent medium do not efficiently trigger star formation.

A setup that can be studied in a more controlled fashion than a turbulent cloud is the triggering of star formation in an individual core. When the ionization front hits the core, it may compress it so heavily that gravitational collapse is triggered. On the other hand, ionization heats up the material and can photoevaporate the core. The remaining material of this competition may form low-mass stars (Hester & Desch 2005) or brown dwarfs (Whitworth & Zinnecker 2004).

The interaction of a shock with a dense clump is called "cloud-crushing". The cloud-crushing scenario has been studied numerically both for supernova shocks and ionization fronts. The triggering of gravitational collapse by an ionization front is called "radiation-driven implosion" (Sandford et al. 1982). The first extensive studies of the fate of the shocked cloud showed that strong vortex rings can be produced (Klein et al. 1994). The mixing properties of the cloud depend sensitively on the initial density distribution (Nakamura et al. 2006). Furthermore, simulations of dense clumps exposed to an ionizing flux but without strong shocks show the generation of kinetic energy (Kessel-Deynet & Burkert 2003) and fragmentation of the clump (Esquivel & Raga 2007). The high temperatures on one side of the clump give rise to the "rocket effect", which accelerates the gas away from the source of radiation (Oort & Spitzer 1955). This is because the cold gas at the surface of the clump facing the star becomes heated. Thus, it expands into the postshock medium, carrying momentum with it, and consequently the clump accelerates. Radiation-gasdynamical simulations of "cloud-crushing" have also been used to match observations of H II regions, especially in the Eagle Nebula (Williams et al. 2001, Miao et al. 2006). However, these observations can also be modeled by the more realistic dynamical expansion of H II regions into a turbulent molecular cloud (Mellema et al. 2006, Gritschneder et al. 2009b).

---

[6]This means that the radiation just removes the envelopes around newly formed stars, but does not trigger their formation.

2 Physics of Star Formation

> Nicht allein in Rechnungssachen
> Soll der Mensch sich Mühe machen;
> Sondern auch der Weisheit Lehren
> Muß man mit Vergnügen hören.
>
> *(Wilhelm Busch, Max und Moritz)*

Many different physical processes play a role in the formation of stars. This chapter is devoted to present the physical background which is later needed for the numerical simulations. Obviously, a good understanding of the involved physics is an inevitable necessity to perform correct simulations.

The material presented here has its origin in diverse physical fields. First of all, stars form from gas in the interstellar medium, which can be described hydrodynamically (section 2.1). This gas has a heterogeneous composition, consisting of several atomic, molecular and dust species. This gives rise to various heating and cooling processes (section 2.2). Of course, the gas in the molecular cloud core can only collapse because of its mutual gravitational attraction (section 2.3). Finally, stars emit radiation, which in turn interacts with the ambient gas (section 2.4). To successfully model massive star formation, all these different aspects must be taken into account. Moreover, additional physical processes are important to understand and model observations of star forming regions (section 2.5).

2.1 Hydrodynamics

Since stars form from gas in the interstellar medium, simulations of star formation are always of hydrodynamical nature. The Reynolds number in the ISM is generally high enough that viscous effects can be neglected. Thus, the interstellar gas can be modeled as an ideal fluid which is governed by the compressible Euler equations. This section will shortly review the fundamentals of ideal, compressible hydrodynamics. More information can be found in the literature (Clarke & Carswell 2007, Shu 1992, Laney 1998, Landau & Lifschitz 1991, Davidson 2004, Chorin & Marsden 2000, Marsden & Hughes 1994).

2.1.1 Euler Equations

The motion of ideal fluids is governed by the Euler equations. The Euler equations are a set of balance equations for mass, momentum and energy. They are also called

2 Physics of Star Formation

conservation laws because for isolated systems, the total mass, momentum and energy are constant with time.

The conservation of mass is expressed by the continuity equation

$$\partial_t \rho + \nabla \cdot (\rho \boldsymbol{v}) = 0. \tag{2.1}$$

Here, ρ is the density and $\boldsymbol{v}$ the velocity of the fluid. The continuity equation states that the change of mass in a given volume can only be due to mass fluxes through the surface of the volume. Mass is not created or destroyed.

In the absence of external force fields, the only force that acts on a fluid is caused by pressure gradients. Hence, momentum balance leads to the equation

$$\partial_t (\rho \boldsymbol{v}) + \nabla \cdot (\rho \boldsymbol{v} \otimes \boldsymbol{v}) = -\nabla P \tag{2.2}$$

with the pressure P. This is nothing but Newton's second law of motion for a fluid parcel. If the fluid is not isolated anymore, for example in the presence of gravitational fields, additional force terms must be added to the right-hand side of equation (2.2).

If the fluid flow is not isothermal, an energy equation must be added to the system. This accounts for heating and cooling of the fluid by adiabatic compression and expansion, respectively. It reads

$$\partial_t (\rho e_{\text{tot}}) + \nabla \cdot [(\rho e_{\text{tot}} + P) \boldsymbol{v}] = 0, \tag{2.3}$$

where the total specific energy e_{tot} is the sum of the specific internal energy e and the specific kinetic energy of the gas,

$$e_{\text{tot}} = e + \frac{1}{2} \boldsymbol{v}^2. \tag{2.4}$$

The specific energies e_{tot} and e are related to the energy densities ϵ_{tot} and ϵ through the relations $\epsilon_{\text{tot}} = \rho e_{\text{tot}}$ and $\epsilon = \rho e$. Additional heating and cooling processes like photoionization heating and molecular line cooling can be included by adding source terms to the right-hand side of equation (2.3).

The Euler equations form a system of partial differential equations,

$$\partial_t \rho + \nabla \cdot (\rho \boldsymbol{v}) = 0, \tag{2.5}$$
$$\partial_t (\rho \boldsymbol{v}) + \nabla \cdot (\rho \boldsymbol{v} \otimes \boldsymbol{v}) + \nabla P = 0, \tag{2.6}$$
$$\partial_t (\rho e_{\text{tot}}) + \nabla \cdot [(\rho e_{\text{tot}} + P) \boldsymbol{v}] = 0. \tag{2.7}$$

We count five equations and six variables. An additional equation is needed to close this system of equations. We use the equation of state of an ideal gas, as introduced in section 2.1.2. It relates the thermal pressure P to the internal energy e of the gas.

Additional source terms must be added to the Euler equations (2.5)–(2.7) to model massive star formation. Of course, gravity is important for the gas to collapse. We present the Euler equations with gravity included in section 2.3. Heating and cooling processes in the interstellar gas and dust are also important. We discuss these processes and the respective source terms in section 2.2.

For numerical reasons, it is sometimes disadvantageous to advect the total specific energy e_{tot} via equation (2.3). This can be the case if the kinetic energy is much larger than the internal energy as in strong shocks or rarefaction waves. Then the calculation of the specific internal energy e via equation (2.4) can lead to cancelation errors and even negative values for e. To avoid this, one can directly advect e instead of e_{tot} with the equation

$$\partial_t(\rho e) + \nabla \cdot [(\rho e + P)\boldsymbol{v}] - \boldsymbol{v} \cdot \nabla P = 0 \tag{2.8}$$

and then calculate e_{tot} from e and $\boldsymbol{v}$. This method is numerically more robust.

2.1.2 Thermodynamics of Ideal Gases

We assume that we can model the interstellar medium as an ideal gas. The thermal equation of state of an ideal gas of N particles in a volume V reads

$$P = \frac{Nk_\text{B}T}{V}, \tag{2.9}$$

where P is the pressure and T the temperature of the gas. With the number density

$$n = \frac{N}{V}, \tag{2.10}$$

this can be written as

$$P = nk_\text{B}T. \tag{2.11}$$

The mass density ρ and the number density n are related by the equation

$$\rho = \mu m_\text{p} n \tag{2.12}$$

when the mean molecular weight μ is given in multiples of the proton mass m_p. Since molecular clouds consist mainly of molecular hydrogen H_2 and some small amounts of metals, $\mu \gtrsim 2$ for the ISM. We can now formulate the thermal equation of state (2.11) in terms of mass density as

$$P = \frac{\rho k_\text{B} T}{\mu m_\text{p}}. \tag{2.13}$$

With the gas constant R, this takes the form

$$P = \frac{\rho R T}{\mu}. \tag{2.14}$$

Since we model the ISM as a molecular ideal gas, we have to think about the possibility of the excitation of the internal degrees of freedom of these molecules. It turns out that we can model a gas of H_2 molecules as monatomic until the temperature reaches a few hundred Kelvin (Commerçon et al. 2008). In the simulations, these high temperatures virtually only appear in H II regions, where the hydrogen is ionized anyway and again the gas can be considered monatomic[1].

---

[1] More details on the effects of ionization on the thermodynamics can be found in section 2.4.5.

Hence, we can assume that the ISM is a calorically perfect gas, which means that the specific heat at constant volume

$$c_V = \left(\frac{\partial e}{\partial T}\right)_{V,N} \tag{2.15}$$

and the specific heat at constant pressure

$$c_P = \left(\frac{\partial h}{\partial T}\right)_{P,N} \tag{2.16}$$

are constant. Here we have introduced the specific enthalpy

$$h = e + \frac{P}{\rho}. \tag{2.17}$$

We can write equations (2.15) and (2.16) in the form

$$e = c_V T \tag{2.18}$$

and

$$h = c_P T, \tag{2.19}$$

respectively. The adiabatic index

$$\gamma = \frac{c_P}{c_V} \tag{2.20}$$

of a molecular ideal gas is related to the degrees of freedom f of a single molecule by

$$\gamma = \frac{f+2}{f}. \tag{2.21}$$

For a monatomic gas with $f = 3$, we find $\gamma = 5/3$. We also have the relation

$$c_P - c_V = \frac{R}{\mu} \tag{2.22}$$

with the ideal gas constant R and the mean molecular weight μ. Hence, we can represent the specific heats c_V and c_P in terms of R, μ and the adiabatic index (2.20) as

$$c_V = \frac{R}{(\gamma - 1)\mu} \tag{2.23}$$

and

$$c_P = \frac{\gamma R}{(\gamma - 1)\mu}, \tag{2.24}$$

respectively. With equations (2.18) and (2.23), the equation of state (2.14) takes the form

$$P = (\gamma - 1)\rho e. \tag{2.25}$$

This relation is frequently used in the simulations to calculate the pressure P from the specific energy e and vice versa. If we combine equation (2.25) with equation (2.14) we find that the conversion between specific internal energy and temperature is given by

$$T = \frac{(\gamma - 1)\mu e}{R}. \tag{2.26}$$

An ideal gas with $\gamma = 5/3$ undergoing gravitational collapse heats up adiabatically. In section 2.2, we discuss various cooling processes which dissipate the heat such that the collapse is effectively isothermal over a wide range in density. In a very simplified approach, one can model these cooling processes by setting $\gamma \gtrsim 1$. This adiabatic index corresponds to gas particles with a very large number of internal degrees of freedom, which can absorb the energy such that the temperature stays constant.

2.1.3 Sound Waves and Shocks

Linear perturbation theory shows that in a fluid at rest, small perturbations travel at a velocity

$$c_\mathrm{s} = \sqrt{\left(\frac{\partial P}{\partial \rho}\right)_s}, \tag{2.27}$$

where the derivative must be taken[2] at constant specific entropy[3] s. This velocity is called the sound speed. At constant entropy, the expression PV^γ is constant, and so is $P\rho^{-\gamma}$. Hence, we can write the pressure as

$$P = P_0 \left(\frac{\rho}{\rho_0}\right)^\gamma. \tag{2.28}$$

The sound speed is then given by

$$c_\mathrm{s} = \sqrt{\frac{\partial P}{\partial \rho}} = \sqrt{\gamma \frac{P}{\rho}}, \tag{2.29}$$

or with eqation (2.13) by

$$c_\mathrm{s} = \sqrt{\frac{\gamma k_\mathrm{B} T}{\mu m_\mathrm{p}}}. \tag{2.30}$$

For the cold ISM, typical values are $T \approx 20\,\mathrm{K}$ and $\mu \approx 2$, so that the sound speed is $c_\mathrm{s} \approx 0.4\,\mathrm{km/s}$. Inside an H II region however, we have $T \approx 10^4\,\mathrm{K}$ and $\mu \approx 1$, which leads to $c_\mathrm{s} \approx 11.7\,\mathrm{km/s}$.

Nonlinear perturbations can propagate much faster than the sound speed. The Mach number

$$\mathcal{M} = \frac{v}{c_\mathrm{s}} \tag{2.31}$$

---

[2] Strictly speaking, we would have to distinguish between the adiabatic and the isothermal sound speed. We will not do this here since they only differ by a factor of order unity.
[3] For isolated, ideal fluids, the specific entropy is conserved along trajectories of fluid particles.

relates the actual velocity v of the fluid to the sound speed c_s. A flow with $\mathcal{M} < 1$ is called subsonic, whereas a flow with $\mathcal{M} > 1$ is called supersonic. Typical Mach numbers in the ISM have values of $\mathcal{M} = 0.1$ to $\mathcal{M} = 10$ (Elmegreen & Scalo 2004).

Equation (2.29) with $P \propto \rho^\gamma$ shows that the sound speed is higher in denser regions. The density along a nonlinear perturbation is not constant, so that the gas in the denser regions propagates faster than in the underdense regions. This leads to a steepening of the wave front and in the end to the formation of a discontinuity. At this shock, the Euler equations in differential form no longer hold, and an integral form must be used[4].

2.2 Heating and Cooling

The Euler equations (2.5)–(2.7) describe a single, isolated fluid. The ISM, however, consists of atoms and molecules of different species, which chemically react with each other, emit and absorb photons, and generate or dissipate heat (Dyson & Williams 1980, Spitzer 1978). We account for these processes by adding source terms to the energy equation, describing the input of energy to and depletion of energy from the fluid. The set of Euler equations including source terms reads

$$\partial_t \rho + \nabla \cdot (\rho \boldsymbol{v}) = 0, \tag{2.32}$$
$$\partial_t (\rho \boldsymbol{v}) + \nabla \cdot (\rho \boldsymbol{v} \otimes \boldsymbol{v}) + \nabla P = 0, \tag{2.33}$$
$$\partial_t (\rho e_{\text{tot}}) + \nabla \cdot [(\rho e_{\text{tot}} + P)\boldsymbol{v}] = \Gamma - \Lambda. \tag{2.34}$$

The first new term is the heating rate

$$\Gamma = \Gamma_{\text{ph}} + \Gamma_{\text{st}} + \Gamma_{\text{acc}}. \tag{2.35}$$

It consists of the photoionization heating rate Γ_{ph}, the stellar heating rate Γ_{st} and the accretion heating rate Γ_{acc}. The photoionization heating rate is caused by photons with high energy which can ionize atomic hydrogen and thereby transfer kinetic energy to the photoelectrons. The other two terms stem from the interaction of non-ionizing radiation with interstellar dust. Non-ionizing radiation may still be strong enough to release electrons from the surface of dust grains by the photoelectric effect. The electron carries the energy difference between the photon energy and the work function as kinetic energy and again heats up the gas. The concrete functional form of these heating terms is given in section 2.4.4.

The second new term is the cooling rate

$$\Lambda = \Lambda_{\text{ml}} + \Lambda_{\text{mol}} + \Lambda_{\text{gd}}. \tag{2.36}$$

Here, contributions arise from the metal line cooling Λ_{ml}, the molecular line cooling Λ_{mol} and the cooling by interstellar dust Λ_{gd}. With metal line cooling we summarize

---

[4]The finite volume methods used in this work all solve the integral form of the Euler equations numerically, but nethertheless, special care must be taken when shocks are present.

the cooling curve of Dalgarno & McCray (1972). They considered cooling processes in the ionized ISM and found that the relevant processes are collisional cooling for a partly ionized gas and bremsstrahlung for a fully ionized gas. The collisional cooling consists of several components, such as the electron impact excitation of electronic and fine structure levels of neutral and ionized atoms. Since metals dominate the collisional cooling, we denote this cooling function as metal line cooling. It is important to counteract the photoionization heating and has a lower temperature threshold of 100 K. The other two cooling processes are relevant for the protostellar collapse phase. The molecular line cooling is taken from data of Neufeld et al. (1995). The cooling table represents the radiative cooling rate of the most important molecules in the ISM. Molecules have internal degrees of freedom, namely rotational, vibrational and electronic excitation levels. These levels can be collisionally excited and decay again by emission of photons. In this way, the energy is carried away from the gas. Similar considerations hold for the excitation of dust grains. The cooling by gas-dust interaction is taken into account following Goldsmith (2001). In this model, gas and dust particles are separate species with their own temperatures, but they are coupled via collisions. We first determine the dust equilibrium temperature from the balance

$$\Lambda_{\text{gd}} - \Lambda_{\text{dust}} = 0, \tag{2.37}$$

where Λ_{dust} represents the cooling of gas particles by black-body radiation. The dust temperature then determines the strength of the gas-dust coupling term. The details of this method can be found in Banerjee et al. (2006).

2.3 Gravity

The collapse of interstellar gas to protostars is properly described by the equations of Newtonian gravity. After a short review of the relevant equations in section 2.3.1, we describe the problem of gravitational instability in section 2.3.2, which leads to the Jeans condition. In hydrostatic equilibrium, self-gravitating gas assembles in form of Bonnor-Ebert spheres, which we discuss in section 2.3.3. Finally, in section 2.3.4 we describe the typical formation process of a protostar.

2.3.1 Poisson Equation and Source Terms

The relation between the gas density ρ and its gravitational potential ϕ is given by the Poisson equation

$$\Delta \phi = 4\pi G \rho, \tag{2.38}$$

where G is Newton's constant. Gradients in the gravitational potential give rise to an acceleration

$$\boldsymbol{g} = -\nabla \phi. \tag{2.39}$$

This acceleration enters the momentum equation (2.2), so that the momentum equation in the presence of gravitational forces reads

$$\partial_t(\rho \boldsymbol{v}) + \nabla \cdot (\rho \boldsymbol{v} \otimes \boldsymbol{v}) + \nabla P = \rho \boldsymbol{g}. \tag{2.40}$$

2 Physics of Star Formation

A corresponding term represents the energy input by the gravitational field, which modifies the energy equation (2.3) to

$$\partial_t(\rho e_{\text{tot}}) + \nabla \cdot [(\rho e_{\text{tot}} + P)\boldsymbol{v}] = \rho \boldsymbol{v} \cdot \boldsymbol{g}. \tag{2.41}$$

Thus, the full set of Euler equations with gravity read

$$\partial_t \rho + \nabla \cdot (\rho \boldsymbol{v}) = 0, \tag{2.42}$$
$$\partial_t(\rho \boldsymbol{v}) + \nabla \cdot (\rho \boldsymbol{v} \otimes \boldsymbol{v}) + \nabla P = \rho \boldsymbol{g}, \tag{2.43}$$
$$\partial_t(\rho e_{\text{tot}}) + \nabla \cdot [(\rho e_{\text{tot}} + P)\boldsymbol{v}] = \rho \boldsymbol{v} \cdot \boldsymbol{g}. \tag{2.44}$$

2.3.2 Gravitational Instability and Jeans Criterion

The onset of gravitational instability in a homogeneous medium was first investigated by Jeans (1902). He used a linear perturbation analysis to study the growth of perturbations around a hydrostatic equilibrium solution with constant density and pressure. This analysis is mathematically dubious because it makes use of the famous "Jeans swindle" (Binney & Tremaine 2008, Shu 1992, Zahn 1976): It starts from an equilibrium state which does not exist. However, the result of Jeans can also be derived in a more rigorous fashion as a suitable limiting process (Kiessling 2003, Chu 2007), which may explain why it is in good agreement with physical reality. The result of the linear stability analysis is that modes with wave vector $\boldsymbol{k}$ obey the dispersion relation

$$\omega^2 = c_s^2 k^2 - 4\pi G \rho_0, \tag{2.45}$$

where ρ_0 is the background density. This equation shows that modes become unstable if the wave number k is smaller than the critical Jeans wave number

$$k_J = \sqrt{\frac{4\pi G \rho_0}{c_s^2}}. \tag{2.46}$$

This wave number corresponds to a Jeans length of

$$\lambda_J = \frac{2\pi}{k_J} = \sqrt{\frac{\pi c_s^2}{G \rho_0}}. \tag{2.47}$$

Perturbations exceeding this size are gravitationally unstable. This result can also be interpreted in terms of a critical mass scale. A sphere with a diameter of a Jeans length has a mass of

$$M_J = \frac{4\pi}{3} \rho_0 \left(\frac{\lambda_J}{2}\right)^3 = \frac{\pi^{5/2}}{6G^{3/2}} \frac{c_s^3}{\rho_0^{1/2}}. \tag{2.48}$$

If the perturbation is more massive than M_J, it will collapse. For isothermal gas, equation (2.30) shows that the Jeans mass scales as $M_J \sim \rho_0^{-1/2} T^{3/2}$. The larger the temperature and the smaller the density, the larger the Jeans mass.

2.3 Gravity

2.3.3 Bonnor-Ebert Spheres

Bonnor-Ebert spheres are isothermal spheres of gas in hydrostatic equilibrium. In hydrostatic equilibrium, $v = 0$, the momentum equation (2.40) reduces to

$$\frac{\nabla P}{\rho} = -\nabla \phi. \tag{2.49}$$

If we take the divergence of this equation, we can insert the Poisson equation (2.38) to eliminate the gravitational potential ϕ,

$$\nabla \cdot \left(\frac{\nabla P}{\rho}\right) = -4\pi G \rho. \tag{2.50}$$

In spherical symmetry and in an isothermal medium with $P = c_s^2 \rho$, this equation simplifies to the Lame-Emden equation

$$\frac{1}{r^2}\frac{d}{dr}\left(\frac{r^2}{\rho}\frac{d\rho}{dr}\right) = -\frac{4\pi G}{c_s^2}\rho. \tag{2.51}$$

It can be made dimensionless with a typical length scale

$$r_0 = \frac{c_s}{\sqrt{4\pi G \rho_0}} \tag{2.52}$$

and the core density ρ_0. By setting

$$\xi = \frac{r}{r_0} \tag{2.53}$$

and

$$\rho(\xi) = \rho_0 \exp(-\Phi(\xi)), \tag{2.54}$$

we arrive at the standard dimensionless form

$$\frac{1}{\xi^2}\frac{d}{d\xi}\left(\xi^2 \frac{d\Phi}{d\xi}\right) = \exp(-\Phi). \tag{2.55}$$

This is a second order ordinary differential equation, which can be integrated numerically if appropriate initial conditions are specified. The first initial condition

$$\Phi(0) = 0 \tag{2.56}$$

follows from the definition of the core density, $\rho(0) = \rho_0$. The second condition stems from the expectation of the hydrostatic core to be flat, or pressure gradients to be absent, respectively,

$$\left.\frac{d\rho}{dr}\right|_{r=0} = 0, \tag{2.57}$$

which translates into

$$\left.\frac{d\Phi}{d\xi}\right|_{\xi=0} = 0. \tag{2.58}$$

2 Physics of Star Formation

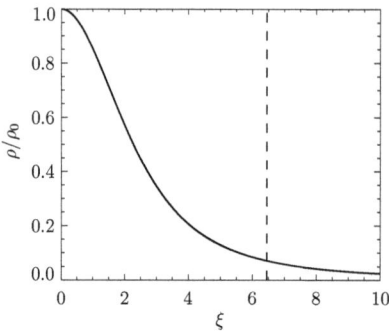

Figure 2.1: Density profile of a Bonnor-Ebert sphere. The critical radius $\xi_c \approx 6.45$ is marked by the dashed line. The densiy profile drops nearly like ξ^{-2}.

The resulting density profile is plotted in figure 2.1. It declines with radius roughly as ξ^{-2}.

Although the numerical solution can be extended to an arbitrarily large radius, Bonnor (1956) and Ebert (1955) have shown by thermodynamic arguments that the sphere becomes gravitationally unstable if it is too large. The critical radius turns out to be $\xi_c \approx 6.45$. A comparison of the length scale (2.52) with the Jeans length (2.47) for a density ρ_0 shows that $\lambda_J = 2\pi r_0$ or $\lambda_J \approx r_c$. The Bonnor-Ebert mass, which is the mass contained inside the sphere of critical radius r_c, turns out to contain 1.5 Jeans masses (Banerjee et al. 2004).

For the simulations, it is important that the parameters of the Bonnor-Ebert sphere are chosen such that the sphere is in pressure equilibrium with its environment. For example, one can choose freely the size of the sphere via the radius ξ_{bnd} and the temperatures inside the sphere and in the environment. The temperatures can be converted into an external pressure p_{ext} and a sound speed inside the sphere $c_{s,sph}$. For the sphere to be in pressure equilibrium, the condition

$$c_{s,sph}^2 \rho_{bnd} = p_{ext} \tag{2.59}$$

for the density at the boundary ρ_{bnd} must hold. Following equation (2.54), the core density is then given by the relation

$$\rho_0 = \rho_{bnd} \exp(\Phi(\xi_{bnd})). \tag{2.60}$$

Now every property of the Bonnor-Ebert sphere is determined.

Bonnor-Ebert spheres have been observed in molecular clouds, the most spectacular example being the dark cloud Barnard 68 (Alves et al. 2001).

2.3.4 Gravitational Collapse of Self-Gravitating Gas

We will now quickly review how gravitational collapse proceeds (Tohline 1982, Stahler & Palla 2004). For the simple case of a sphere with homogeneous density ρ, the problem of isothermal pressure-free collapse can be analytically solved (Hunter 1962). It turns out that the whole sphere collapses within a free-fall time

$$t_\mathrm{ff} = \sqrt{\frac{3\pi}{32G\rho}}. \tag{2.61}$$

The free-fall time only depends on the initial density, not on the temperature. Hence, all points of the sphere have the same density at a given time and arrive at the center simultaneously. The collapse is called homologous.

If the initial sphere is not homogeneous in density, like the Bonnor-Ebert spheres described above, then the collapse will be non-homologous. Here, the denser, inner parts will collapse quicker than the less dense, outer parts because their free-fall time is shorter. The collapse proceeds inside-out.

The problem of gravitational collapse with pressure gradients taken into account is much more complicated. Various self-similar solutions to this problem are known (Larson 1969, Penston 1969, Shu 1977, Hunter 1977). Historically, the Larson-Penston and Shu solutions attracted the most attention. They differ in the initial conditions and lead to different profiles for density and velocity as well as different protostellar accretion rates. They are related to different ideas about the physics involved in protostellar collapse (Klessen 2003). Here we just remark that numerical simulations favor the Larson-Penston solution (Foster & Chevalier 1993) and that a collapsing isothermal sphere will generally approach a density profile $\rho \propto r^{-2}$ (Larson 2003). This is illustrated in figure 2.2, where simulation results of spherically-symmetric isothermal collapse for two different initial density profiles are shown.

A more realistic assumption than spherical symmetry is that the molecular cloud core also has some initial rotation. The conservation of angular momentum then leads to the formation of an accretion disk. As outlined in section 1.1.1, the formation of stars by accretion from a disk has culminated in the standard picture of star formation (Shu et al. 1987). A useful quantity for rotating cores is the ratio of rotational to gravitational energy

$$\beta = \frac{E_\mathrm{rot}}{E_\mathrm{grav}} \tag{2.62}$$

with

$$E_\mathrm{rot} = \frac{1}{2} \int_V \rho v^2 \, dV \tag{2.63}$$

and

$$E_\mathrm{grav} = -\frac{1}{2} \int_V \rho \phi \, dV. \tag{2.64}$$

Here, the core is assumed to be in pure rotational motion, so that the kinetic energy is equal to the rotational energy. This is applicable if the core is set up in solid-body rotation, which is often done in simulations.

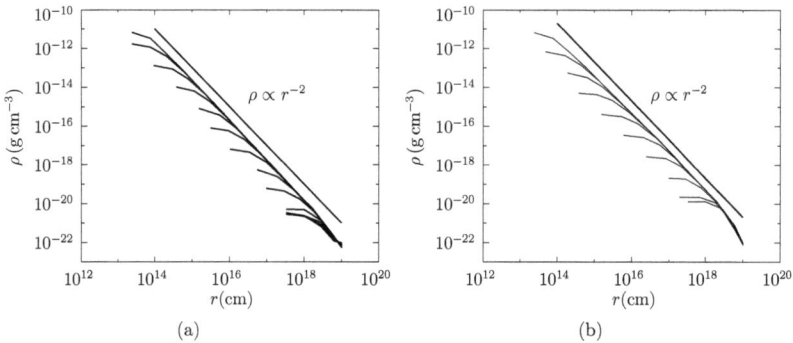

Figure 2.2: Spherically-symmetric isothermal collapse. (a) Collapse of a Bonnor-Ebert sphere with $\xi = 10$ and a mass of $172 M_\odot$. (b) Collapse of a sphere with flat inner profile of $0.5\,\mathrm{pc}$ radius and a $\rho \propto r^{-3/2}$ fall-off with a total mass of $1000 M_\odot$. The time span between the successive radial profile plots steadily decreases during the collapse. The straight lines show profiles with $\rho \propto r^{-2}$ fall-off.

2.4 Radiation

In this section, we review the basic notions and concepts of radiative transfer and radiation hydrodynamics. We begin with the physical description of radiation fields and consider the propagation of radiation through a medium. We then specialize to the raytracing approximation used in the simulations. Finally, we consider the interaction of non-ionizing and ionizing radiation with a fluid and discuss the expansion of ionization fronts. More information can be found in the literature on radiation physics (Mihalas & Mihalas 1984, Peraiah 2002, Rybicki & Lightman 1979, Castor 2004, Shu 1991) and the interstellar medium (Osterbrock 1989, Dyson & Williams 1980, Spitzer 1978).

2.4.1 Radiation Fields

We now introduce the basic quantities and relations of radiation transfer. The specific intensity $I(\boldsymbol{x}, t; \boldsymbol{n}, \nu)$ of a radiation field at position $\boldsymbol{x}$ at time t in direction $\boldsymbol{n}$ with frequency ν is defined such that the amount of radiant energy E in the frequency interval $[\nu, \nu + \mathrm{d}\nu]$ transported across an area element $\mathrm{d}A$ into a solid angle $\mathrm{d}\Omega$ around $\boldsymbol{n}$ in a time interval $\mathrm{d}t$ is given by

$$\mathrm{d}E = I(\boldsymbol{x}, t; \boldsymbol{n}, \nu) \cos\alpha \, \mathrm{d}A \, \mathrm{d}t \, \mathrm{d}\Omega \, \mathrm{d}\nu, \qquad (2.65)$$

where α is the angle between the direction $\boldsymbol{n}$ and the normal to the area element $\mathrm{d}A$. From now on we will abbreviate $I(\boldsymbol{x}, t; \boldsymbol{n}, \nu)$ as I_ν. The specific flux F_ν is the amount of energy traversing a unit area within a unit frequency interval per unit time in all directions, that is

$$F_\nu = \oint I_\nu \cos\alpha \, d\Omega. \tag{2.66}$$

We also define the total intensity

$$I = \int_0^\infty I_\nu \, d\nu \tag{2.67}$$

and the total flux

$$F = \int_0^\infty F_\nu \, d\nu. \tag{2.68}$$

Other quantities of interest are the mean specific intensity

$$J_\nu = \frac{1}{4\pi} \oint I_\nu \, d\Omega \tag{2.69}$$

and the mean total intensity

$$J = \frac{1}{4\pi} \oint I \, d\Omega. \tag{2.70}$$

We now consider the energy content of the radiation field. It is related to the photon number density $\psi(\boldsymbol{x},t;\boldsymbol{n},\nu)$ of photons in the frequency interval $[\nu, \nu + d\nu]$ traveling into a solid angle $d\Omega$ around the direction $\boldsymbol{n}$. Since photons travel at the speed of light c, all photons in a volume $dA\,c\,dt$ will have left this volume in the time interval dt. The number of photons that are transported across a surface dA into the solid angle $d\Omega$ is then given by $\psi \cos\alpha \, dA\, c\, dt\, d\Omega\, d\nu$ if the normal to dA is subtending the direction $\boldsymbol{n}$ at an angle α. The photon energy is $h\nu$, where h is Planck's constant, so the total energy transported by the photons is

$$dE = ch\nu\psi(\boldsymbol{x},t;\boldsymbol{n},\nu)\cos\alpha \, dA\, dt\, d\Omega\, d\nu. \tag{2.71}$$

A comparison with the definition of the specific intensity (2.65) yields the relation

$$I_\nu = ch\nu\psi_\nu. \tag{2.72}$$

The specific energy density $u_\nu(\boldsymbol{x},t;\nu)$ can be written in terms of the specific photon number density as

$$u_\nu = h\nu \oint \psi_\nu \, d\Omega. \tag{2.73}$$

With the relation (2.72) and the definition of the mean specific intensity (2.69) we find

$$u_\nu = \frac{4\pi}{c} J_\nu. \tag{2.74}$$

Hence, the total energy density of the radiation field is

$$u = \frac{4\pi}{c} J. \tag{2.75}$$

2.4.2 Radiative Transfer

Since energy is a conserved quantity, any change in the radiation energy along a ray must be caused by either emission or absorbtion of radiation along this ray. We consider a ray with length ds and cross sectional area dA. We can relate the change of radiant energy entering the ray at position $\boldsymbol{x}$ within a solid angle $d\Omega$ around the direction $\boldsymbol{n}$ normal to the surface area dA at time t and leaving the ray at position $\boldsymbol{x} + d\boldsymbol{x}$ at time $t + dt$ in the corresponding cone in terms of the intensity (2.65) as

$$\bigl(I(\boldsymbol{x} + d\boldsymbol{x}, t + dt; \boldsymbol{n}, \nu) - I(\boldsymbol{x}, t; \boldsymbol{n}, \nu)\bigr)\, dA\, dt\, d\Omega\, d\nu$$
$$= \bigl(j(\boldsymbol{x}, t; \boldsymbol{n}, \nu) - \alpha(\boldsymbol{x}, t; \boldsymbol{n}, \nu) I(\boldsymbol{x}, t; \boldsymbol{n}, \nu)\bigr)\, ds\, dA\, dt\, d\Omega\, d\nu, \quad (2.76)$$

where j_ν is the emission coefficient and α_ν the absorption coefficient. With the relation $ds = c\, dt$ we find

$$I(\boldsymbol{x} + d\boldsymbol{x}, t + dt; \boldsymbol{n}, \nu) - I(\boldsymbol{x}, t; \boldsymbol{n}, \nu) = \left(\frac{1}{c}\frac{\partial I_\nu}{\partial t} + \frac{\partial I_\nu}{\partial s}\right) ds \quad (2.77)$$

and thus

$$\frac{1}{c}\frac{\partial I_\nu}{\partial t} + \frac{\partial I_\nu}{\partial s} = j_\nu - \alpha_\nu I_\nu. \quad (2.78)$$

This is the equation of radiative transfer. The derivative with respect to the path length s depends on the coordinate system. In Cartesian coordinates, we get

$$\frac{\partial I_\nu}{\partial s} = \frac{\partial x}{\partial s}\frac{\partial I_\nu}{\partial x} + \frac{\partial y}{\partial s}\frac{\partial I_\nu}{\partial y} + \frac{\partial z}{\partial s}\frac{\partial I_\nu}{\partial z} = (\boldsymbol{n}\cdot\nabla)I_\nu. \quad (2.79)$$

In spherical coordinates, this expression is more complicated, but it simplifies again under the assumption of spherical symmetry. We model the star emitting the radiation as a point source and place it in the origin of the coordinate system. Then each ray is a coordinate line of constant ϑ, φ, the path length s is identical to the radial coordinate r and the ray direction $\boldsymbol{n}$ is identical to the radial direction $\boldsymbol{r}/r$. For this simple case,

$$\frac{\partial I_\nu}{\partial s} = \frac{\partial I_\nu}{\partial r}. \quad (2.80)$$

Furthermore, we neglect the time derivative and the emissivity j_ν and assume that the absorption coefficient α_ν is isotropic, that is it does not depend on the direction vector $\boldsymbol{n}$. Then the radiative transfer equation reads

$$\frac{dI_\nu}{dr} = -\alpha_\nu I_\nu. \quad (2.81)$$

We can also look at this equation from a microscopic viewpoint. The medium contains absorbers with a number density n and a cross section σ_ν, so that the loss of energy is given by

$$dI_\nu\, dA\, d\Omega\, dt\, d\nu = -I_\nu(n\sigma_\nu\, dA\, dr)\, d\Omega\, dt\, d\nu \quad (2.82)$$

or
$$dI_\nu = -n\sigma_\nu I_\nu \, dr. \tag{2.83}$$

By comparing equations (2.81) and (2.83), we find for the absorption coefficient

$$\alpha_\nu = n\sigma_\nu. \tag{2.84}$$

We can express the absorption coefficient in terms of the mass density ρ as well. Then we have

$$\alpha_\nu = \rho \kappa_\nu \tag{2.85}$$

with the opacity coefficient κ_ν.

We can directly integrate equation (2.81) to

$$I_\nu(r) = I_\nu(0) \exp\left(-\int_0^r \alpha_\nu(r') \, dr'\right). \tag{2.86}$$

With the optical depth

$$\tau_\nu(r) = \int_0^r \alpha_\nu(r') \, dr' \tag{2.87}$$

we can write this as

$$I_\nu(r) = I_\nu(0) \exp(-\tau_\nu(r)). \tag{2.88}$$

This equation shows that the intensity falls off by a factor of $1/e$ if the optical depth reaches the value $\tau_\nu = 1$. Material with $\tau_\nu < 1$ is called optically thin, material with $\tau_\nu > 1$ is called optically thick. Basically, a photon of frequency ν can travel through an optically thin medium without being absorbed, whereas it almost cannot pass an optically thick medium without absorption. If α_ν is spatially constant, then we simply have

$$I_\nu(r) = I_\nu(0) \exp(-\alpha_\nu r). \tag{2.89}$$

If the absorption cross section σ_ν is spatially constant, it is helpful to introduce the column density

$$N(r) = \int_0^r n(r') \, dr', \tag{2.90}$$

We can then rewrite the optical depth $\tau_\nu(r)$ with help of the relation (2.84) and equation (2.87) as

$$\tau_\nu(r) = \sigma_\nu N(r). \tag{2.91}$$

The absorption cross section of atomic hydrogen can be taken from Osterbrock (1989). If ν_T is the frequency at ionization threshold, corresponding to the ionization energy of hydrogen, 13.6 eV, it has the form

$$\sigma_\nu = \sigma_T \left[\beta \left(\frac{\nu}{\nu_T}\right)^{-s} + (1-\beta)\left(\frac{\nu}{\nu_T}\right)^{-s-1}\right] \tag{2.92}$$

for $\nu > \nu_\mathrm{T}$, with the parameters $\beta = 1.34$, $\sigma_\mathrm{T} = 6.3 \cdot 10^{-18}\,\mathrm{cm}^2$ and $s = 2.99$. Since it is independent of temperature, we can use equation (2.91) to calculate the optical depth for a gas containing atomic hydrogen only with

$$N(r) = \int_0^r n_\mathrm{HI}(r')\,\mathrm{d}r', \tag{2.93}$$

where n_HI is the number density of atomic hydrogen. We also define the number density of ionized hydrogen n_HII and the total number density of hydrogen

$$n_\mathrm{H} = n_\mathrm{HI} + n_\mathrm{HII}. \tag{2.94}$$

With the ionization fractions

$$x_\mathrm{HI} = \frac{n_\mathrm{HI}}{n_\mathrm{H}}, \qquad x_\mathrm{HII} = \frac{n_\mathrm{HII}}{n_\mathrm{H}} \tag{2.95}$$

we can express the column density (2.93) in terms of the total number density n_H as

$$N(r) = \int_0^r n_\mathrm{H}(r')x_\mathrm{HI}(r')\,\mathrm{d}r'. \tag{2.96}$$

The opacity coefficient of dust is not independent of temperature. Hence, we cannot use column densities, but we must use equation (2.87) to compute the optical depth of a gas-dust mixture,

$$\tau_\nu(r) = \int_0^r \rho(r')\kappa_\nu(T(r'))\,\mathrm{d}r'. \tag{2.97}$$

Since our radiative transfer is not frequency-dependent, we must use some integrated mean opacities. According to Mihalas & Mihalas (1984), the Planck mean opacity κ_P is the best to use as a heating term in the Euler equation. Let

$$B_\nu(T) = \frac{2h}{c^2}\frac{\nu^3}{\exp(h\nu/k_\mathrm{B}T) - 1} \tag{2.98}$$

be the specific intensity of a black body of temperature T, then the Planck mean opacity is defined by the relation

$$\int_0^\infty \kappa_\nu(T,\rho)B_\nu(T)\,\mathrm{d}\nu = \kappa_\mathrm{P}(T,\rho)\int_0^\infty B_\nu(T)\,\mathrm{d}\nu, \tag{2.99}$$

which gives

$$\kappa_\mathrm{P}(T,\rho) = \frac{\pi}{\sigma T^4}\int_0^\infty \kappa_\nu(T,\rho)B_\nu(T)\,\mathrm{d}\nu. \tag{2.100}$$

Here, we have used the Stefan-Boltzmann law

$$\int_0^\infty B_\nu(T)\,\mathrm{d}\nu = \frac{\sigma T^4}{\pi}. \tag{2.101}$$

with the Stefan-Boltzmann constant σ.

The choice of suitable opacities for non-ionizing radiation is very difficult. First, the opacities must cover the whole density and temperature regime that occurs in the simulations. Opacities derived for stellar physics or non-ionized protoplanetary disks are not sufficient in this respect. Secondly, even the most state-of-the-art data from Ferguson et al. (2005) and Semenov et al. (2003) do not agree at all in a range between 10^3 K and 10^4 K, which may have an influence on the non-ionizing heating feedback. Given the large uncertainties in the present data and their inexpedience for our problem, we choose to take the simple approach of Krumholz et al. (2007a) and parametrize the opacities of Pollack et al. (1994) as

$$\kappa_\mathrm{P}(T) = \begin{cases} 0.3 + 7.0(T/375) & T \leq 375, \\ 7.3 + 0.7(T-375)/200 & 375 < T \leq 575, \\ 3.0 + 0.1(T-575)/100 & 575 < T \leq 675, \\ 2.8 + 0.3(T-675)/285 & 675 < T \leq 960, \\ 3.1 - 3.0(T-960)/140 & 960 < T \leq 1100, \\ 0.1 & T > 1100, \end{cases} \tag{2.102}$$

where T is given in K and κ_P in cm^{2}g^{-1}. This at least allows a straigtforward comparison of the simulation results.

2.4.3 Point Sources

In the simulations, we neglect any kind of diffuse radiation and consider only the radiation emitted by the protostar, which we model as a point source[5]. It can be shown (Rybicki & Lightman 1979) that the specific flux at a distance r from a star of radius r_star with a uniform specific intensity at its surface $I_\nu(0)$ is given by[6]

$$F_\nu(r) = \pi I_\nu(r)\left(\frac{r_\mathrm{star}}{r}\right)^2. \tag{2.103}$$

Because of the symmetry of the problem, there is only a flux in radial direction. Since all radiation is originating from a point source, the integral (2.66) evaluated at any point other than the source will have contributions only along the line of sight to the

---

[5] For the dimensions considered here, which range from the scale of some pc $\approx 3 \cdot 10^{18}$ cm to several AU $\approx 1.5 \cdot 10^{13}$ cm, the protostar with a radius of a few solar radii $R_\odot = 6.96 \cdot 10^{10}$ cm can be assumed to be pointlike as far as radiative transfer is concerned.

[6] Here, we neglect the minute difference in the optical depth for rays emanating from different points of the surface, that is the effect of limb darkening (Peraiah 2002).

2 Physics of Star Formation

source, for which $\cos\alpha = 1$ holds. Thus, the integrals (2.66) and (2.69) are equal, and we find

$$F_\nu(r) = 4\pi J_\nu(r). \tag{2.104}$$

We can use formula (2.88) as a solution of the radiative transfer equation along rays,

$$I_\nu(r) = I_\nu(0)\exp(-\tau_\nu(r)). \tag{2.105}$$

If we model the specific intensity of the protostar with a black-body spectrum of temperature T_{star}, that is

$$I_\nu(0) = B_\nu(T_{\text{star}}), \tag{2.106}$$

we can analytically express the mean specific intensity of the point source as

$$J_\nu(r) = \left(\frac{r_{\text{star}}}{r}\right)^2 \frac{1}{2c^2} \frac{h\nu^3}{\exp(h\nu/k_B T_{\text{star}}) - 1} \exp(-\tau_\nu(r)). \tag{2.107}$$

Using equation (2.74), we also have an analytical form for the specific energy density of the radiation field,

$$u_\nu(r) = \left(\frac{r_{\text{star}}}{r}\right)^2 \frac{2\pi}{c^3} \frac{h\nu^3}{\exp(h\nu/k_B T_{\text{star}}) - 1} \exp(-\tau_\nu(r)). \tag{2.108}$$

The spectrum of the protostar is totally determined by its luminosity L, that is the total amount of energy emitted by the star per unit time, and its effective temperature T_{star}. We can find the radius r_{star} of the star with the Stefan-Boltzmann law

$$L = 4\pi\sigma r_{\text{star}}^2 T_{\text{star}}^4. \tag{2.109}$$

Usually, the luminosity is given in multiples of the solar luminosity, which is $L_\odot = 3.862 \cdot 10^{33}\,\text{erg}\,\text{s}^{-1}$. We take the values for the luminosity L and effective temperature T_{star} of the protostar as function of mass from a zero-age main sequence (ZAMS) for a solar metallicity of $Z = 0.02$, which was generated from the freely available stellar evolution code EZ (see Paxton (2004)). Figure 2.3(a) shows the Hertzsprung-Russell diagram of this main sequence, which was calculated for masses between 0.1 and $100 M_\odot$. The mass dependence of luminosity and temperature is plotted in figure 2.3(b).

2.4.4 Coupling to Hydrodynamics

In our model, we neglect the effects of radiation pressure. Radiation pressure will lead to the formation of radiation bubbles as described by Krumholz et al. (2005a). However, these bubbles show up only when the star has already reached $17 M_\odot$, and at this point the effect of ionization feedback is dominant. Thus, we only couple the radiation field to the energy equation in form of ionization heating and dust heating terms. Krumholz (2006) and Krumholz et al. (2007a) showed that dust heating is

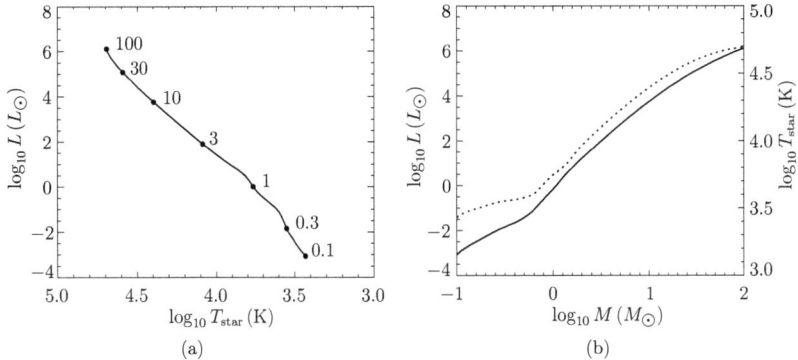

Figure 2.3: Zero-age main sequence of the protostellar model. (a) Hertzsprung-Russell diagram. The dots indicate stellar masses for selected data points between 0.1 and 100$M_\odot$. (b) Luminosity (solid) and temperature (dotted) as function of stellar mass.

important to suppress fragmentation of massive cores and accretion disks in massive star formation.

According to Osterbrock (1989), the photoionization heating rate is given by

$$\Gamma_{\rm ph} = n_{\rm HI} \int_{\nu_{\rm T}}^{\infty} \frac{4\pi J_\nu}{h\nu} \sigma_\nu h(\nu - \nu_{\rm T}) \, d\nu, \qquad (2.110)$$

where σ_ν is taken from equation (2.92). This equation can easily be interpreted physically. By definition, the term $4\pi J_\nu$ represents the energy transported by the radiation field per unit area per unit time and per unit frequency interval, so that this term divided by $h\nu$ is the corresponding number of photons. This photon number times the cross section σ_ν is the number of ionizations per hydrogen atom per second per frequency interval, and the term $h(\nu - \nu_{\rm T})$ represents the excess energy of the photon, which is assumed be fully converted into kinetic energy of the photoelectron and thus constitutes a heat source for the gas. Hence, the integral represents the energy input by ionization per hydrogen atom per second, and $\Gamma_{\rm ph}$ is the total energy input per second per unit volume. With the specific mean intensity (2.107), this expression becomes

$$\Gamma_{\rm ph}(r) = n_{\rm HI} \left(\frac{r_{\rm star}}{r}\right)^2 \frac{2\pi h}{c^2} \int_{\nu_{\rm T}}^{\infty} \frac{\sigma_\nu \nu^2 (\nu - \nu_{\rm T}) \exp(-\tau_\nu(r))}{\exp(h\nu/k_{\rm B} T_{\rm star}) - 1} \, d\nu \qquad (2.111)$$

for the ray under consideration. Here, the optical depth $\tau_\nu(r)$ is determined by equation (2.91).

2 Physics of Star Formation

Following Krumholz et al. (2007b), the dust heating term can be written to lowest order in v/c as

$$\Gamma_{\rm d} = \kappa_{\rm P}\rho c u, \qquad (2.112)$$

which can be calculated even simpler since the dust opacities (2.102) are gray. More specifically, we consider the stellar heating rate $\Gamma_{\rm st}$, which is caused by the luminosity of the protostar itself. The total energy density follows from equations (2.108) as

$$u(r) = \left(\frac{r_{\rm star}}{r}\right)^2 \frac{\pi}{c} \exp(-\tau(r)) \int_0^\infty \frac{2h\nu^3/c^2}{\exp(h\nu/k_{\rm B}T_{\rm star}) - 1} d\nu. \qquad (2.113)$$

Because the integrand is the intensity of a black body, the Stefan-Boltzmann law (2.101) leads to

$$u(r) = \left(\frac{r_{\rm star}}{r}\right)^2 \frac{\sigma}{c} \exp(-\tau(r)) T_{\rm star}^4. \qquad (2.114)$$

So the stellar heating term at position r on the ray is given by

$$\Gamma_{\rm st}(r) = \sigma \left(\frac{r_{\rm star}}{r}\right)^2 \kappa_{\rm P}(T(r))\rho(r) \exp(-\tau(r)) T_{\rm star}^4. \qquad (2.115)$$

We also include a simple model of accretion heating (Frank et al. 1992). We use the formula

$$L_{\rm acc} = G\frac{M\dot{M}}{r_{\rm acc}} \qquad (2.116)$$

to determine the accretion luminosity $L_{\rm acc}$ from the mass of the protostar M and the accretion rate $\dot{M}$. The idea behind this model is that the potential energy of the gas which is released during accretion is fully converted into radiation at the accretion radius $r_{\rm acc}$. To define the accretion radius, we use the prestellar evolution model of Hosokawa & Omukai (2009). We interpolate their data to find $r_{\rm acc}$ as a function of M and $\dot{M}$. It turns out that the accretion radius can vary from a few solar radii to more than $100R_\odot$, depending on the accretion rate and the protostellar mass. We assume that this radiation has the form of a black-body spectrum, so that we can determine the corresponding temperature $T_{\rm acc}$ from the Stefan-Boltzmann law. This simply adds another heating term of the form (2.115) to the energy equation,

$$\Gamma_{\rm acc}(r) = \sigma \left(\frac{r_{\rm acc}}{r}\right)^2 \kappa_{\rm P}(T(r))\rho(r) \exp(-\tau(r)) T_{\rm acc}^4. \qquad (2.117)$$

2.4.5 Ionization

The time evolution of the ionization fractions (2.95) is governed by a rate equation. This equation accounts for neutral hydrogen atoms which become ionized and ionized hydrogen atoms which catch an electron and recombine to form a neutral hydrogen atom. The rate equation reads

$$\frac{{\rm d}x_{\rm HII}}{{\rm d}t} = x_{\rm HI}(A_{\rm p} + A_{\rm c}) - x_{\rm HII}n_e\alpha_{\rm R}. \qquad (2.118)$$

We assume that there is always a small fraction of free electrons from carbon atoms in the interstellar medium, which are easily singly ionized by the interstellar UV field. Thus, we set

$$n_\text{e} = n_\text{H\,II} + n_\text{C} \tag{2.119}$$

with $n_\text{C} = 7.1 \cdot 10^{-7} n_\text{H}$.

The first contribution to the ionization rate is the photoionization rate (Osterbrock 1989)

$$A_\text{p} = \int_{\nu_\text{T}}^{\infty} \frac{4\pi J_\nu}{h\nu} \sigma_\nu \, d\nu. \tag{2.120}$$

In the previous section, we have identified the integrand as the number of ionizations per hydrogen atom per second per frequency interval, so that the integral over all frequencies yields the desired rate. With the specific mean intensity (2.107), the photoionization rate becomes

$$A_\text{p}(r) = \left(\frac{r_\text{star}}{r}\right)^2 \frac{2\pi}{c^2} \int_{\nu_0}^{\infty} \frac{\sigma_\nu \nu^2 \exp(-\tau_\nu(r))}{\exp(h\nu/k_\text{B}T_\text{star}) - 1} \, d\nu. \tag{2.121}$$

The second contribution is the collisional ionization rate (Cox & Tucker 1969)

$$A_\text{c} = A_\text{H\,I} n_\text{e} \sqrt{T} \exp(-I_\text{H\,I}/k_\text{B}T) \tag{2.122}$$

with $A_\text{H\,I} = 5.84 \cdot 10^{-11} \, \text{cm}^3 \, \text{K}^{-1/2} \, \text{s}^{-1}$ and the hydrogen ionization energy $I_\text{H\,I} = 13.6 \, \text{eV}$.

When an electron and a proton recombine, the hydrogen atom can be in any possible bound state. As the electron cascades down to the ground state, photons will be emitted. When the electron was not directly captured in the ground state, these photons will not be energetic enough to ionize further hydrogen atoms, so that they can escape from the H II region unhindered. Otherwise, the secondary photons will contribute to the ionization rate. The on-the-spot approximation (Osterbrock 1989) effectively accounts for these secondary ionizing photons by assuming that they are absorbed close to their source. Recombinations to the ground state are not considered since they lead to ionization close-by. The recombination coefficient which incorporates only recombinations to excited states is commonly called "Case B". Using this approximation, the radiative recombination rate is given by (Rijkhorst et al. 2006)

$$\alpha_\text{R} = \alpha_\text{R}(10^4 \, \text{K}) \left(\frac{T}{10^4 \, \text{K}}\right)^{-0.7} \tag{2.123}$$

with $\alpha_\text{R}(10^4 \, \text{K}) = 2.59 \cdot 10^{-13} \, \text{cm}^3 \, \text{s}^{-1}$.

Since the ionization changes the number of free particles of the gas, we need to modify the equation of state. The total number density is now $n = n_\text{H} + n_\text{e}$, so that the equation of state (2.11) takes the form

$$P = \frac{(n_\text{H} + n_\text{e})k_\text{B}T}{\mu}. \tag{2.124}$$

Note the appearance of the mean molecular weight μ in the denominator. This is because the quantity n in equation (2.11) is the number density of free particles. In the case of molecular clouds, this is roughly the number of H_2 molecules. Hence, the number density of hydrogen atoms n_H overcounts the number density of free particles n_{H_2} by a factor of $\mu \approx 2$, that is

$$n_H = \mu n_{H_2}. \tag{2.125}$$

Thus, we have to introduce the factor μ if we want to write equation (2.11) in terms of n_H. However, equation (2.124) holds strictly only in the non-ionized regions. We model the effects of ionization on thermodynamics by keeping μ constant and introducing the additional number density n_e. In the extreme of full ionization, we have $n_e \approx n_H$, and the number density which contributes to equation (2.11) is underestimated by a factor of 2. A more consistent way would be a true multi-fluid treatment with a chemical network and variable μ, but this is beyond the scope of the present work. Our method assures that the thermodynamics of neutral and partially ionized gas is connected smoothly.

Using $n_{H\,II} \approx n_e$ and relation (2.125), we can write equation (2.124) in the form

$$P = n_{H_2}(1 + x_{H\,II})k_B T \tag{2.126}$$

or, with equation (2.12) and $n = n_{H_2}$, as

$$P = \frac{\rho R T}{\mu}(1 + x_{H\,II}). \tag{2.127}$$

This is the modified form of equation (2.14). By the same token, the factor $(1 + x_{H\,II})$ is introduced in the caloric equation of state, so that c_V and c_P now satisfy

$$c_P - c_V = \frac{R}{\mu}(1 + x_{H\,II}) \tag{2.128}$$

and thus

$$c_V = \frac{R}{(\gamma - 1)\mu}(1 + x_{H\,II}). \tag{2.129}$$

Hence, the factors $(1 + x_{H\,II})$ in equation (2.127) and in

$$e = \frac{R}{(\gamma - 1)\mu}(1 + x_{H\,II})T \tag{2.130}$$

cancel, and the relation

$$P = (\gamma - 1)\rho e \tag{2.131}$$

still holds. If we combine this result with equation (2.127), we find

$$T = \frac{(\gamma - 1)\mu}{R} \frac{e}{1 + x_{H\,II}}, \tag{2.132}$$

which is the modified form of equation (2.26). This completes the set of thermodynamic relations for the ionized gas.

2.4 Radiation

We now argue that we can treat the radiative transfer for the ionizing and non-ionizing radiation separately. Ionizing radiation will heat the gas only as long as the medium is optically thin. This is because the exponential factor in the ionization heating rate (2.111) and the photoionization rate (2.121) becomes very small otherwise. As long as this is not the case, the gas is being ionized very effectively to an ionization fraction of $x_{\mathrm{HII}} \approx 1$. But equation (2.96) shows that this means $\tau \approx 0$ in the ionized material, whereas τ grows very rapidly if $x_{\mathrm{HII}} \approx 0$. Indeed, it turns out that there is a rather sharp transition between the ionized material for which $\tau < 1$ and the non-ionized gas with $\tau > 1$. This will be demonstrated in section 4.1. We conclude that the ionizing radiation by far dominates the gas heating in the ionized region, whereas it is totally negligible in the non-ionized domain.

2.4.6 Expansion of H II Regions and Ionization Fronts

When a massive star forms, the surrounding neutral gas from which it formed gets increasingly ionized. The expansion of this H II region into the ambient ISM can be roughly divided into two stages. Initially, the photon flux is so high that the ionization front propagates highly supersonically into the neutral gas. The gas heats up to 10^4 K, but since the expansion timescale is much smaller than the dynamical timescale, movement of the gas can be neglected in this phase. Ionization fronts in this stage are called R-type.

The radius R_S at which this expansion stops can be estimated by equating the number N of emitted ionizing photons per second with the number of recombinations per second in the H II region,

$$N = \frac{4}{3}\pi R_S^3 \alpha_R n_{\mathrm{HII}} n_e. \tag{2.133}$$

Solving for R_S, we get

$$R_S = \left(\frac{3N}{4\pi \alpha_R n_{\mathrm{HII}} n_e}\right)^{1/3}. \tag{2.134}$$

This radius is called the Strömgren radius (Strömgren 1939). This estimate of course assumes that the density inside the H II region is constant and does not change with time, which is both wrong. Nethertheless, it gives a useful length scale.

After the Strömgren radius is reached, all ionizing photons are depleted, but the ionized gas has a two orders of magnitude higher pressure than the neutral gas. The ionization front now enters the pressure-driven second stage of its expansion and becomes a D-type front. The gas dynamics cannot be neglected anymore. In fact, a shock front between the hot, ionized and the cold, neutral gas develops. Spitzer (1978) found that this ionization shock expands into a homogeneous medium approximately as

$$R(t) = R_S \left(1 + \frac{7}{4}\frac{c_s t}{R_S}\right)^{4/7}, \tag{2.135}$$

where c_s is the sound speed in the ionized gas. This analytical result is important because it is one of the few dynamical predictions numerical treatments of ionization can be tested against (see section 4.1).

The transition between the ionized and the neutral gas is very sharp, but it is not smooth. Initial perturbations in the shock surface will grow due to an imbalance of the ram pressure in the ionized gas and the thermal pressure in the neutral gas (Giuliani 1979, Vishniac 1983). This thin-shell instability leads to the growth of finger-like structures in the shock front.

Another instability is the shadowing instability (Williams 1999). Here, dense clumps shield ionizing radiation in R-type ionization fronts, so that the front corrugates and the parts of the front where the radiation was not shielded make the transition to D-type earlier. This can again trigger the formation of fingers. A stellar wind can have a similar effect (Freyer et al. 2003, 2006).

Ionization fronts become even more unstable when they expand into a medium with a density gradient (Tenorio-Tagle 1979, Franco et al. 1990, García-Segura & Franco 1996, García-Segura et al. 2006, Arthur & Hoare 2006, Whalen & Norman 2008a,b). To summarize, the Spitzer solution (2.135) gives a crude idea of how ionization fronts expand, but the detailed dynamics at the ionization front can be much more complicated.

2.5 Synthetic Observations

To compare our simulation results with observations of massive star forming regions, we wish to generate synthetic observations of our data. H II regions around massive stars can easily be observed with radio telescopes. The source of emission at cm wavelengths is mainly free-free radiation, while sub-mm wavelenghts are dominated by dust emission (Gordon & Sorochenko 2002). Since free-free radiation is more straightforward to model than dust emission, we focus on observations at cm wavelengths. We discuss the generation of synthetic radio continuum maps in section 2.5.1. The velocity structure of the accretion flow can be traced by line emission, both from radio recombination lines and from molecular lines. This is described in section 2.5.2.

2.5.1 Free-free Radiation

Accelerated electric charges emit electromagnetic radiation. In particular, free electrons in the plasma of an H II region are scattered by hydrogen ions and emit thermal bremsstrahlung (Padmanabhan 2000). Since the electron is free before and after the emission of the photon, bremsstrahlung is a free-free emission process. The inverse process of free-free emission is free-free absorption. We call the emission and absorption coefficient for free-free radiation j_ν^{ff} and α_ν^{ff}, respectively. Since both emission and absorption now appear on an extended region, we have to slightly modify the radiative transfer analysis from the previous section, which assumed that emission only results from a point source.

2.5 Synthetic Observations

The time-independent version of the radiative transfer equation (2.78) for the case of free-free radiation reads

$$\frac{dI_\nu}{dr} = j_\nu^{\text{ff}} - \alpha_\nu^{\text{ff}} I_\nu. \tag{2.136}$$

Here we have used the fact that scattering of radiation can generally be neglected (Kraus 1966, Gordon & Sorochenko 2002). To solve this differential equation, we again introduce the optical depth

$$\tau_\nu^{\text{ff}}(r) = \int_0^r \alpha_\nu^{\text{ff}}(r') \, dr'. \tag{2.137}$$

After dividing by α_ν^{ff}, equation (2.136) can be written in terms of τ_ν^{ff} as

$$\frac{dI_\nu}{d\tau_\nu^{\text{ff}}} = S_\nu - I_\nu. \tag{2.138}$$

The expression

$$S_\nu = \frac{j_\nu^{\text{ff}}}{\alpha_\nu^{\text{ff}}} \tag{2.139}$$

is called the source function. For notational simplicity, we will suppress the superscript ff from now on. With the integrating factor e^{τ_ν}, we find

$$\frac{d}{d\tau_\nu}\left(e^{\tau_\nu} I_\nu\right) = e^{\tau_\nu} I_\nu + e^{\tau_\nu} \frac{dI_\nu}{d\tau_\nu} = e^{\tau_\nu} S_\nu \tag{2.140}$$

and thus

$$I_\nu(\tau_\nu) = I_\nu(0) e^{-\tau_\nu} + e^{-\tau_\nu} \int_0^{\tau_\nu} e^{\tau_\nu'} S_\nu(\tau_\nu') \, d\tau_\nu'. \tag{2.141}$$

When the region is not illuminated from behind, we have $I_\nu(0) = 0$. Furthermore, by Kirchhoff's law, we have

$$S_\nu = B_\nu(T) \tag{2.142}$$

with the specific intensity of a black body B_ν of temperature T. The wavelengths under consideration are well within the Rayleigh-Jeans limit $h\nu \ll k_B T$, so that

$$B_\nu(T) = \frac{2\nu^2}{c^2} k_B T. \tag{2.143}$$

This leads to the integral for the brightness temperature

$$T(\tau_\nu) = e^{-\tau_\nu} \int_0^{\tau_\nu} e^{\tau_\nu'} T(\tau_\nu') \, d\tau_\nu'. \tag{2.144}$$

This integral can be readily computed for a given domain once the optical depth (2.137) is known. To this end, we need to calculate the absorption coefficient for free-free radiation. This computation turns out to be a complicated matter, and various

2 Physics of Star Formation

VLA band	wavelength (cm)	beam width (arcsec)	sensitivity (mJy)
4	400	24	160
P	90	6.0	4.0
L	20	1.4	0.061
C	6.0	0.4	0.058
X	3.6	0.24	0.049
U	2.0	0.14	1.0
K	1.3	0.08	0.11
Q	0.7	0.05	0.27

Table 2.1: VLA telescope parameters. The VLA can observe at eight bands with wavelengths from 400 cm to 0.7 cm. The beam width is given in terms of the full width at half maximum (FWHM) in arcsec, the sensitivity in mJy.

approximations exist for different regimes. Gordon & Sorochenko (2002) give the formula

$$\alpha_\nu = 0.212 \left(\frac{n_e}{\text{cm}^{-3}}\right) \left(\frac{n_i}{\text{cm}^{-3}}\right) \left(\frac{T_e}{\text{K}}\right)^{-1.35} \left(\frac{\nu}{\text{Hz}}\right)^{-2.1} \text{cm}^{-1} \quad (2.145)$$

for thermal bremsstrahlung from H II regions, with the number density of free electrons n_e and hydrogen ions n_i as well as the electron temperature T_e. We set $n_i = n_e$ and assume that T_e is equal to the gas temperature T in thermal equilibrium.

To generate synthetic free-free emission images from the simulation data, we calculate the integral (2.144) across H II regions. This is easy to do if the direction of integration is parallel to the coordinate axes of the grid. For an arbitrary direction, a full raytracing procedure is necessary. To generate images from arbitrary angles, we have implemented free-free emission into the radiative transfer code RADMC-3D and checked the two codes against each other.

To simulate an observation, we must convert the brightness temperature to a flux density[7] via

$$F_\lambda = \frac{2k_B T}{\lambda^2} \Omega_S. \quad (2.146)$$

Here, Ω_S is the solid angle of the telescope beam. Afterwards, the resulting image is convolved with the beam width, and appropriate noise is added to the image following the algorithm described in Mac Low et al. (1991). The telescope parameters used for the synthetic observations of the Very Large Array (VLA) are listed in table 2.1.

When images of the same region are calculated for different wavelengths, the spectral energy distribution (SED) can be obtained. To get some intuition for how the SED should look like, we assume a homogeneous density and temperature distribution inside the H II region. Then the integral (2.144) is simply

$$T(\tau_\nu) = T(1 - e^{-\tau_\nu}). \quad (2.147)$$

---

[7]Radio astronomers measure the flux density in Jy (Jansky). The conversion from cgs units to Jy is $1\,\text{Jy} = 10^{-23}\,\text{erg}\,\text{s}^{-1}\,\text{cm}^{-2}\,\text{Hz}^{-1}$.

2.5 Synthetic Observations

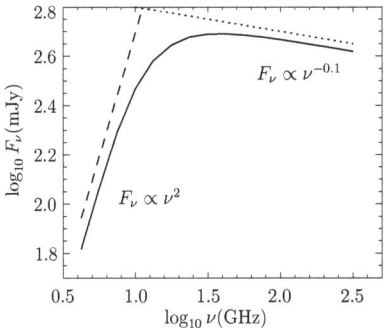

Figure 2.4: SED of a homogeneous cylinder. The cylinder is oriented parallel to the direction of integration, so that all rays across the cylinder have the same optical depth. The dashed line has a slope $F_\nu \propto \nu^2$, the dotted line $F_\nu \propto \nu^{-0.1}$. The SED shows that the cylinder is optically thick below 10 GHz and optically thin above 100 GHz.

For an optically thin medium ($\tau_\nu \ll 1$), we have

$$T(\tau_\nu) = T\tau_\nu \propto \nu^{-2.1}, \qquad (2.148)$$

and with equation (2.146) we find the scaling for the flux density

$$F_\nu \propto \nu^{-0.1}. \qquad (2.149)$$

If the medium is optically thick ($\tau_\nu \gg 1$), we get

$$T(\tau_\nu) = T \propto \nu^0, \qquad (2.150)$$

and thus

$$F_\nu \propto \nu^2. \qquad (2.151)$$

The optical depth decreases with $\nu^{-2.1}$, so that for low frequencies, when the medium is optically thick, the SED should grow with ν^2, and for higher frequencies, when the medium becomes optically thin, the SED will decline as $\nu^{-0.1}$. This behavior is illustrated in figure 2.4.

2.5.2 Line Emission

When a free electron in the plasma is caught by a proton, a photon is emitted. Usually, the electron does not reside in the ground state immediately, but cascades downwards, successively emitting further photons. A bit confusingly, these bound-bound transitions are called recombination lines, and many of these lines fall into the radio regime

(Rybicki & Lightman 1979, Gordon & Sorochenko 2002). The hydrogen recombination lines are denoted as H$n\alpha$, H$n\beta$, ... for the transitions $n+1 \rightarrow n$, $n+2 \rightarrow n$, ... Since the lines are Doppler-shifted in moving gas, the analysis of radio recombination lines can be used to infer kinematic information about the ionized gas.

A good tracer for the molecular gas is ammonia, NH$_3$. We use a constant NH$_3$ abundance along with H$_2$, but since we do not have molecular hydrogen as a species in the simulation data, we have to get the H$_2$ number density by postprocessing. To this end, we assume that an equilibrium state exists between H$_2$ formation on dust grains and the two collisional dissociation processes H$_2$ + H $\rightarrow$ H + H + H and H$_2$+H$_2$ $\rightarrow$ H$_2$+H+H. The dust formation rate χ_{form} is taken from Hollenbach & McKee (1979), for the two destruction processes $\chi_{\text{dest},1}$ and $\chi_{\text{dest},2}$ we use the interpolation formula from Glover & Mac Low (2007). It interpolates between the low-density (Lepp & Shull 1987, Shapiro & Kang 1987) and high-density (Mac Low & Shull 1986, Martin et al. 1998) limits of these rates. In equilibrium, the respective number densities satisfy the equation

$$\chi_{\text{form}}(n_{\text{H}} + 2n_{\text{H}_2})n_{\text{H}} = \chi_{\text{dest},1}\, n_{\text{H}_2}n_{\text{H}} + \chi_{\text{dest},2}\, n_{\text{H}_2}n_{\text{H}_2}. \tag{2.152}$$

This equation is quadratic in $y = n_{\text{H}_2}/n_{\text{H}}$ and solved by the expression

$$y = -\frac{1}{2}\frac{\chi_{\text{dest},1} - 2\chi_{\text{form}}}{\chi_{\text{dest},2}} + \sqrt{\frac{1}{4}\frac{(\chi_{\text{dest},1} - 2\chi_{\text{form}})^2}{\chi_{\text{dest},2}^2} + \frac{\chi_{\text{form}}}{\chi_{\text{dest},2}}}. \tag{2.153}$$

Since
$$n_{\text{H}} + 2n_{\text{H}_2} = n_{\text{p}}, \tag{2.154}$$

we have
$$n_{\text{H}_2} = \frac{yn_{\text{p}}}{2y+1} \tag{2.155}$$

with the proton number density n_{p}.

The ammonia molecule NH$_3$ exhibits a large number of transitions that can be used to study interstellar gas (Ho & Townes 1983, Stahler & Palla 2004). The molecule has a pyramidal form. The rotational energy of the molecule can be described by two quantum numbers (J, K). In addition, the molecule can undergo inversion transitions where the N atom tunnels trough the barrier of the three H atoms. These transitions are split into several hyperfine structure states. The main hyperfine line of the $(3, 3)$ transition is most useful to study the dynamics of accretion flows because it is not blended with the satellite hyperfine lines (Keto et al. 1987).

We use the radiative transfer code MOLLIE (Keto 1990) to generate the synthetic maps of radio recombination and ammonia lines.

3 Simulations of Star Formation

> Though this be madness, yet there is method in't.
>
> *(William Shakespeare, Hamlet)*

The physics of star formation is too involved to solve the underlying equations analytically. The investigation of three-dimensional problems under realistic conditions demands highly sophisticated numerical simulations (Larson 2007, Klessen et al. 2009). We use the FLASH code, developed by Fryxell et al. (2000) for the simulation of supernova explosions. Since its publication, it has been applied to all kinds of hydrodynamical problems, and it has undergone an ample testing and validation phase (Calder et al. 2002). It was first used for simulations of star formation by Banerjee et al. (2004).

In this chapter, we present the basic methodology of simulations with FLASH. We begin with a general overview on simulation techniques (section 3.1) and the principles of block-structured adaptive mesh refinement (AMR) (section 3.2). We then describe the operator splitting (section 3.3), the hydrodynamics solver (section 3.4), the heating and cooling routines (section 3.5), the gravity solver (section 3.6), the radiation module (section 3.7) and the sink particles (section 3.8). Additional information on can be found in the FLASH manual.

3.1 Overview

The methods of numerical astrophysical hydrodynamics can be crudely divided into three groups: Grid methods (like finite difference, finite volume or spectral methods) take an Eulerian point of view. Space is discretized, and the Euler equations are evaluated on a set of specified points (Laney 1998, LeVeque et al. 1998, Toro 1997, LeVeque 1994). This mesh may be fixed or varying in time, allowing for an adaption to the present situation. Adaptive mesh refinement is particularly beneficial for simulations of star formation because it allows to adequately resolve shocks and collapsing regions while dynamically uninteresting regions do not consume too much computational power.

On the other hand, there are particle methods (most prominently Smoothed Particle Hydrodynamics (SPH), introduced by Lucy (1977) and Gingold & Monaghan (1977)), which are inherently of Lagrangian nature. Here, the fluid is represented by a set of particles, which can be thought of as fluid elements. The Euler equations transform into a set of ordinary differential equations, which can be readily integrated (Monaghan 1992, 2005). SPH automatically has a higher resolution in denser regions because there are more particles, but this also means that underdense regions are not very well resolved.

Thirdly, there are hybrid methods, which try to combine the best of both approaches and overcome their drawbacks. For example, Springel (2009) used an adaptive Lagrangian Voronoi tesselation and solved the equations of hydrodynamics with a Riemann solver. How successful such hybrid methods will be in practice remains to be seen.

The coice of the appropriate method depends on the specific applications under consideration. An advantage of grid methods is that the convergence of numerical schemes is mathematically understood (Laney 1998, LeVeque et al. 1998, Toro 1997, LeVeque 1994), which is not true for SPH. Numerical methods for shocks and magnetohydrodynamics are available for grid codes and well tested. But the artificial introduction of a grid leads to spurious effects like the non-conservation of angular momentum, which makes it less favorable for studies of accretion disks or similar rotating structures.

SPH has generally problems with shocks, hydrodynamic instabilities and magnetic fields, although there is some progress (Agertz et al. 2007, Price 2008, Rosswog & Price 2008). Commerçon et al. (2008) presented a detailed comparison of collapse simulations with SPH and AMR. They found that both methods yield the same results provided that the Jeans mass is sufficiently well resolved.

Present-day simulations of star formation also begin to include radiation feedback. Because the full solution of the radiative transfer problem is so complicated, in practice two major approximations are used: the diffusion approximation and the raytracing approximation. The diffusion approximation is valid if the medium is optically thick, so that the mean free path of the photons is short. The resulting random walk can be considered as a diffusion problem. However, the diffusion approximation becomes bad if the medium is optically thin. A flux limiter must be introduced in order to prevent radiation from propagating faster than the speed of light. This leads to the name flux-limited diffusion for this method (Mihalas & Mihalas 1984, LeVeque et al. 1998). It has been implemented both in grid codes (Turner & Stone 2001, Krumholz et al. 2007b) and in SPH (Whitehouse & Bate 2004, Whitehouse et al. 2005).

The other extreme is the raytracing approximation, which we use. Here, scattering is totally neglected, and radiation only propagates radially from point sources. The neglection of scattering is well justified for the ionizing radiation, whereas for the nonionizing radiation in the dense accretion disk it certainly introduces some error. In particular, the accretion luminosity physically is a result of the gas dynamics at the accretion shock and hence is intrinsically diffuse.

The raytracing technique is frequently used in radiation hydrodynamics for modeling ionizing radiation. An overview of some existing implementations in both grid and SPH codes and their performance in a cosmological setting is given in Iliev et al. (2006, 2009). Other implementations that have recently been used to study ionization feedback in star formation are Bisbas et al. (2009), Raga et al. (2009), Gritschneder et al. (2009a,b), Dale et al. (2005, 2007a,b), Krumholz et al. (2007c), Mac Low et al. (2007), Mellema et al. (2006), Miao et al. (2006), Arthur & Hoare (2006) and Rijkhorst et al. (2006). None of the mentioned methods is perfect. Many lack numerical resolution, an appropriate treatment of the ionization physics or both.

Rijkhorst et al. (2006) have implemented a raytracing technique for FLASH which is very well parallelized and includes a detailed treatment of ionization. We have adapted and extended their code for our purposes and coupled it to sink particles as models of protostars. This allows us to make simulations of star formation with ionization feedback that were not possible so far.

3.2 Adaptive Mesh Technique

Adaptive meshes can be built of structured and unstructured grids. Unstructured grids are usually created by triangulating the computational domain and are heavily used in finite element calculations with irregular boundary geometries. Structured grids consist of a hierarchy of rectangular subgrids, which can be represented by a tree. Since boundaries are unimportant for most astrophysical problems, the restriction to rectangular grids is not a problem, and also curvilinear orthogonal coordinates (like spherical or cylindrical coordinates) can be used in FLASH.

Block-structured adaptive mesh techniques use rectangluar arrays of cells, called blocks, to organize the hierarchy of subgrids. Berger & Oliger (1984) developed this method and Berger & Colella (1989) applied it to gasdynamical problems. Their method is highly flexible and accordingly complicated. De Zeeuw & Powell (1993) presented a simplified algorithm that generates subgrids only by bisection. The FLASH code employs the PARAMESH block-structured AMR library by MacNeice et al. (2000), which incorporates these approaches. It is optimized for simplicity to make the development and maintenance of the parallel code easier.

3.2.1 Data Structure

The structure of a single block is shown in figure 3.1. It consists of $N_{x,b}$ cells in x-direction, $N_{y,b}$ cells in y-direction and $N_{z,b}$ cells in z-direction. The default value is $N_{x,b} = N_{y,b} = N_{z,b} = 8$. In addition to these interior cells of a block, the block contains N_{guard} guardcells at each block boundary. These are used as boundary conditions for the Riemann solver and can be taken either from a neighboring block or from the boundary conditions of the domain. Typically, $N_{\mathrm{guard}} = 4$, because the piecewise-parabolic method needs 4 boundary values.

The blocks are refined by bisecting the block along each direction, thereby creating eight (for three dimensions) new child blocks with twice the resolution of the parent block. This is done in such a way that neigboring blocks differ by at most one level of refinement. After the block is refined, only the child blocks carry the updated data, and the data of the parent block becomes obsolete. This means that at each point of the domain, only the blocks at the highest level of refinement are up-to-date. These blocks are called leaf blocks, because they are the leaves of the hierarchy tree. An example domain with its corresponding hierarchy tree is depicted in figure 3.2.

3 Simulations of Star Formation

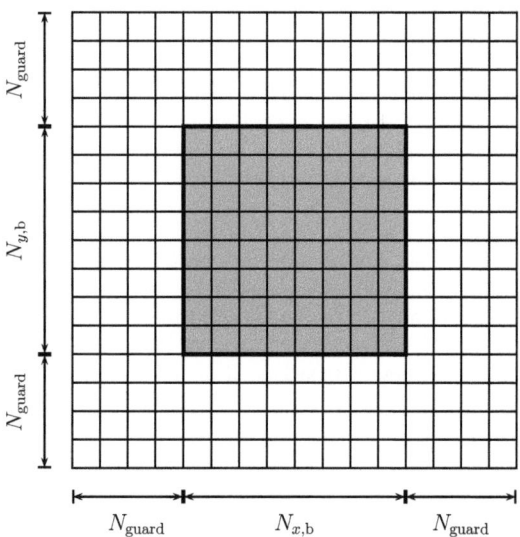

Figure 3.1: A single block with guardcells in two dimensions. By default, each block has $N_{x,b} = N_{y,b} = 8$ cells in each direction and $N_{\text{guard}} = 4$ guardcells that are required by the piecewise-parabolic method as boundary values.

PARAMESH uses a Morton space-filling curve to number the blocks in the domain. The numbers associated with the blocks in figure 3.2 are these Morton numbers. They are chosen as to balance the workload among the different processors and to account for the location of the blocks with respect to each other. For more details on these software engineering aspects of PARAMESH see Calder et al. (2000).

3.2.2 Refinement Criteria

In simulations of star formation, adaptivity is used for two main reaons: For a proper treatment of shocks and to follow the process of gravitational collapse. The shock refinement criterion implemented in FLASH goes back to Löhner (1987). It uses the magnitude of the second derivative relative to the gradient in that cell as an error estimation. For a one-dimensional uniform mesh, it amounts to

$$E_i = \frac{|u_{i+2} - 2u_i + u_{i-2}|}{|u_{i+2} - u_i| + |u_i - u_{i-2}| + \varepsilon\left(|u_{i+2}| + 2|u_i| + |u_{i-2}|\right)}. \quad (3.1)$$

The variable u is the refinement test variable, typically density or pressure to detect shocks, but also ionization fraction to trace ionization fronts. The term in brackets in the denominator filters out small variations in u, with $\varepsilon = 0.01$ by default. In three dimensions, the corresponding expression is

$$E_{i_1,i_2,i_3} = \left[\frac{\sum_{p,q=1}^{3}\left(\frac{\partial^2 u}{\partial x_p \partial x_q}\right)^2}{\sum_{p,q=1}^{3}\left[\frac{1}{2\Delta x_q}\left(\left|\frac{\partial u}{\partial x_p}\right|_{i_q+1} + \left|\frac{\partial u}{\partial x_p}\right|_{i_q-1}\right) + \varepsilon\frac{\overline{|u_{p,q}|}}{\Delta x_p \Delta x_q}\right]^2}\right]^{1/2}, \quad (3.2)$$

where the partial derivatives represent the according finite difference expressions at the cell with index (i_1, i_2, i_3), and $\overline{|u_{p,q}|}$ denotes the average of $|u|$ over neigboring cells in p- and q-direction. A block is refined if for one of its cells E_{i_1,i_2,i_3} is larger than the refinement threshold. It is derefined if E_{i_1,i_2,i_3} is below the derefinement threshold for all its cells.

To trace the collapse of self-gravitating gas, sufficient spatial resolution is crucial to capture the dynamical processes correctly. Truelove et al. (1997) showed that the Jeans length has to be resolved with at least 4 grid cells in order to avoid artificial fragmentation. Banerjee et al. (2004) implemented this criterion in FLASH and pursued the collapse of rotating Bonnor-Ebert spheres through 10 orders of magnitude in density. By allowing less levels of refinement for shocks than for the Jeans condition, this criterion can also be used to detect potentially gravitationally unstable regions in a turbulent medium, because they are then refined more than their environment (Peters et al. 2008).

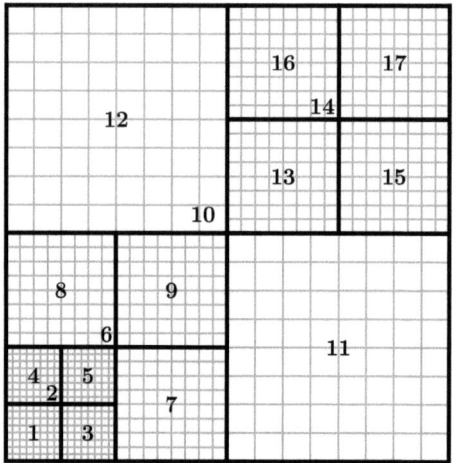
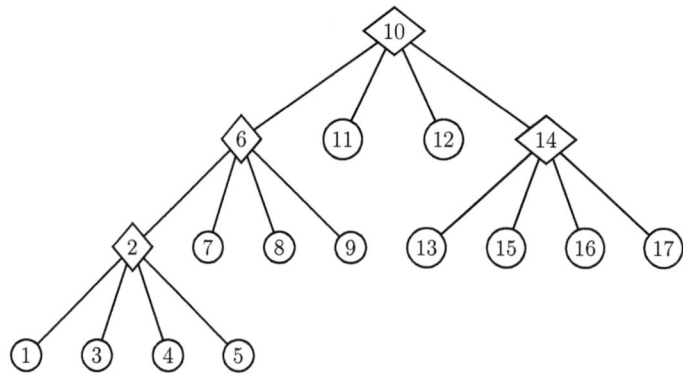

Figure 3.2: A two-dimensional block-structured adaptive mesh with its corresponding hierarchy tree. Each block (black) consists of $N_{x,b} \cdot N_{y,b} = 8^2$ cells (gray). Neighboring blocks differ by at most one level of refinement. The leaf blocks, which carry the updated data, are depicted as circles in the tree diagram (example taken from Fryxell et al. (2000)).

3.2.3 Prolongation and Restriction

When a block is refined, the additional cells have to be filled with data. This process is called prolongation, because the hierarchy tree gets longer. Typically, linear or quadratic interpolation is used for that purpose. However, there may be cases where this leads to unphysical overshoots in the data. To prevent this, it is also possible to just copy the data from a neighboring cell, although this of course reduces the order of accuracy. Interpolation is also used to fill the guard cells if the neighboring block is at a coarser level of refinement.

Derefinement leads to a so-called restriction of the data. The data on the coarser level is obtained by averaging the data on the finer level. Again, this is also done for the guard cells if the neighbor block is on a finer level of refinement.

3.3 Operator Splitting

We cannot solve the full set of Euler equations

$$\partial_t \rho + \nabla \cdot (\rho \boldsymbol{v}) = 0, \tag{3.3}$$

$$\partial_t (\rho \boldsymbol{v}) + \nabla \cdot (\rho \boldsymbol{v} \otimes \boldsymbol{v}) + \nabla P = \rho \boldsymbol{g}, \tag{3.4}$$

$$\partial_t (\rho e_{\text{tot}}) + \nabla \cdot [(\rho e_{\text{tot}} + P) \boldsymbol{v}] = \rho \boldsymbol{v} \cdot \boldsymbol{g} + \Gamma - \Lambda, \tag{3.5}$$

the Poisson equation (2.38), the radiative transfer equation, the ionization rate equation (2.118) and the equation of motion for the sink particles simultaneously. The division of this set of equations into solvable subsets of equations is called operator splitting. The flow chart of the FLASH main cycle is illustrated in figure 3.3.

In the first step, the set of Euler equations (2.42)–(2.44) including gravitational source terms only are solved. Here, dimensional splitting is applied (LeVeque et al. 1998). This means that we treat the problem as one-dimensional, and apply the solution algorithm (usually a Riemann solver, see section 3.4) in only one direction. Then, we solve the one-dimensional problem in another direction, and we do this for all three directions. In this way we solve the three-dimensional problem by solving three one-dimensional problems. Generally, this Godunov splitting is only of first order in time because of the splitting error. This would render it useless to use higher order methods to solve the one-dimensional problems. Strang (1968) found that the order of the method can be increased to second order if the problem is first solved for a half time step for all directions in a given order and then solved in reverse order of directions for another half-step. This method is known as Strang splitting and used in some variant by the FLASH code.

Next, the source terms of the energy equation (2.35) and (2.36) are accounted for. This includes cooling by molecular lines Λ_{mol} and dust Λ_{gd} as well as the non-ionizing radiation heating terms Γ_{st} and Γ_{acc}. We discuss this step in section 3.5. The heating due to the ionizing radiation Γ_{ph} and the metal line cooling Λ_{ml} is done in the following step to self-consistently calculate the ionization fraction.

When the source terms have been treated, the problems related to radiation are considered. This includes the calculation of column densities for each sink particle and adding up the corresponding heating rates $\Gamma_{\rm ph}$, $\Gamma_{\rm st}$ and $\Gamma_{\rm acc}$. When the total photoionization and photoionization heating rates are known, the new ionization fractions and temperatures are calculated. The radiation module is described in section 3.7.

In the next step, the equation of motion of the sink particles is solved, and gas is accreted onto the sink particle. The sink particles move due to gravitational interaction with each other and the gas as well as due to gain of linear momentum by accretion. The mass of the sink particle sets the strength of the stellar heating terms $\Gamma_{\rm ph}$ and $\Gamma_{\rm st}$. The accreted mass per time step yields the accretion rate, which is used to calculate the accretion luminosity and the respective heating term $\Gamma_{\rm acc}$. Sink particles are discussed in section 3.8.

In the last step of one half-cycle, the Poisson equation (2.38) is solved by a multi-grid method. This yields the gravitational potential and acceleration, which are needed as an input for the hydrodynamics solver and the sink particles. Some details of the algorithm can be found in section 3.6.

At the end of one full cycle, the mesh is tested for refinement and derefinement. In order to minimize interpolation errors, it is useful to prevent refinement in successive time steps. Since the gravitational potential is a derived quantity, it is calculated after the refinement at the second half-cycle so that numerical errors are minimized.

3.4 Hydrodynamics

The Euler equations (2.42)–(2.44) belong to the larger class of locally hyperbolic partial differential equations or nonlinear conservation laws (Laney 1998, LeVeque et al. 1998, Toro 1997, LeVeque 1994). There exists a large set of different algorithms to solve these equations, and each has its advantages and disadvantages. We will only give the general idea and then describe the method which we use for our simulations.

As already remarked in section 2.1.1, the Euler equations are balance equations for mass, momentum and energy. Finite volume methods use this fact to solve the integral form of the Euler equations on a grid: The rate of change of mass, momentum or energy in a given cell is equal to the correspondig flux through the cell faces. In a dimensionally split scheme, the equations for mass, one velocity component and energy must be solved simultaneously. Such a multi-dimensional hyperbolic equation can be solved by diagonalizing the Jacobian of the flux function. The eigenvalues then correspond to typical signal velocities in the system, and the total solution can be reconstructed from the solutions for the respective eigenvectors.

The Euler equations for piecewise constant initial data with a single discontinuity can be solved analytically. The solution to this Riemann problem can be applied to two adjacent cells of the grid. Methods that employ this idea are called Riemann solvers. A popular Riemann solver which has proven to be very good in resolving shocks and contact discontinuities is the piecewise parabolic method (PPM) (Colella & Woodward 1984, Woodward & Colella 1984, Colella & Glaz 1985). It is second order

3.4 Hydrodynamics

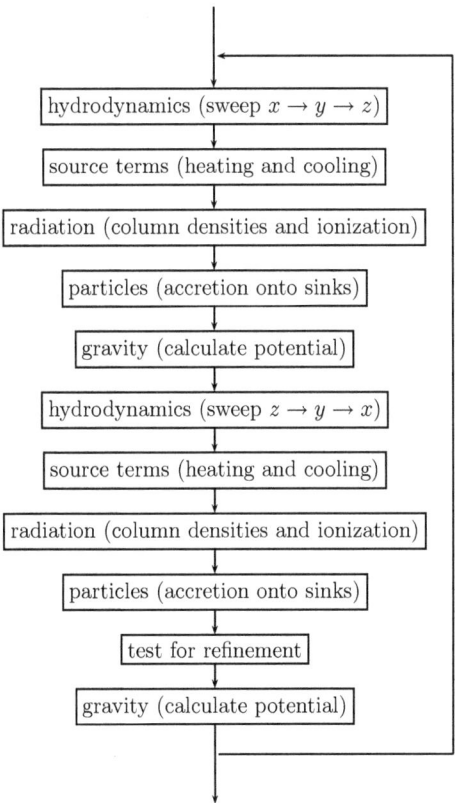

Figure 3.3: Flow chart of the evolution procedure. The coupled set of equations is cut into several pieces, which are handled separately. The operator splitting scheme requires each term to be evaluated twice per time step. After each time step, a check of the refinement criteria can be performed.

3 Simulations of Star Formation

accurate in time and space. Details on the implementation of the piecewise parabolic method in FLASH can be found in Fryxell et al. (2000).

Since PPM is explicit in time, it can become unstable if the time step is too large. The time step should never be larger than allowed by the Courant-Friedrichs-Lewy (CFL) condition

$$\Delta t \max \left(\frac{(|v_x| + c_s)}{\Delta x}, \frac{(|v_y| + c_s)}{\Delta y}, \frac{(|v_z| + c_s)}{\Delta z} \right) < C, \qquad (3.6)$$

where v is the gas velocity and c_s the sound speed in the cell, Δx, Δy and Δz are the cell widths and Δt the time step (Courant et al. 1928). For stability of explicit schemes, $C < 1$ must always hold. This condition amounts to the requirement that no signal in a cell can travel farther than one cell width during one time step. Since FLASH only has a single global time step, the minimum of all local time steps is used for each cell. This causes a slow-down of the simulation when regions with large velocities (shocks), large sound speeds (ionized gas) or small grid spacing (collapsing regions) exist.

3.5 Heating and Cooling

The heating terms (2.35) and cooling terms (2.36) in the energy equation (3.5) are divided into two parts. The ionization heating rate is evaluated together with the ionization fractions in the radiation module. As described in section 3.7.2, the metal line cooling Λ_{ml} is used to counterbalance the ionization heating Γ_{ph}. All the other source terms in equation (3.5) are treated together, as we will discuss now.

The method is an extension to the treatment in Banerjee et al. (2006). We have added the heating terms and a switch to solve the equation implicitly if the change in energy is too large. The first step is to solve equation (2.37) for the dust temperature. We use Newton's method for this purpose. Then also the gas-dust coupling term Λ_{gd} is known. The molecular cooling Λ_{mol} is found with the help of a cooling table. The heating rates Γ_{st} and Γ_{acc} are given analytically by the equations (2.115) and (2.117), respectively. We can then calculate the change in the internal energy density as

$$\Delta \epsilon = (\Gamma_{\mathrm{st}} + \Gamma_{\mathrm{acc}} - \Lambda_{\mathrm{mol}} - \Lambda_{\mathrm{gd}}) \Delta t. \qquad (3.7)$$

If this change is not too large compared to the old value ϵ^i,

$$|\Delta \epsilon| \leq 0.1 \epsilon^i, \qquad (3.8)$$

then we calculate the new internal energy density explicitly,

$$\epsilon^{i+1} = \epsilon^i + \Delta \epsilon. \qquad (3.9)$$

If the condition (3.8) is not satisfied, we take an implicit time step. In any case, we do not allow ϵ to change by more than 30 % in one time step to prevent numerical instabilities. This amounts to solving the equation

$$\epsilon^{i+1} = \epsilon^i + \max \left[\min \left[\left(\Gamma_{\mathrm{st}} + \Gamma_{\mathrm{acc}} - \Lambda_{\mathrm{mol}} \left(\epsilon^{i+1} \right) - \Lambda_{\mathrm{gd}} \left(\epsilon^{i+1} \right) \right) \Delta t, 0.3 \cdot \epsilon^i \right], -0.3 \cdot \epsilon^i \right] \qquad (3.10)$$

for ϵ^{i+1}. To this end, we employ Brent's root-finding method Press et al. (1986).

Since this model does not include important physical processes like heating by cosmic rays, we have to set a temperature floor to prevent the gas from cooling down to 0 K. This temperature floor is typically $T_{\min} = 30$ K. In cases where the simulation develops such high densities that the Jeans length would no longer be resolved at the given temperature and the dynamical formation of sink particles (see section 3.8) is not desired, this temperature floor can be increased to the value

$$T_{\min} = \frac{G\mu m_{\mathrm{p}}}{\pi k_{\mathrm{B}}} \rho (n\Delta x)^2. \tag{3.11}$$

As dicussed in section 3.2.2, $n \geq 4$ must hold to prevent artificial fragmentation. With this dynamical temperature floor, which we call Jeans heating, it is assured that the Jeans length is always resolved, no matter how dense the gas becomes.

3.6 Gravity

Gravity can be incorporated into simulations of star formation in two different ways. The gravitational field can be externally specified or it can be be calculated from the mass density field of the gas. The first method can be useful to model the action of sink particles, while the latter is important for collapse problems.

3.6.1 External Fields

FLASH provides several types of external gravitational fields that define a gravitational acceleration (2.39) without reference to a gravitational potential, so that the gas reacts only passively. Interesting for simulations of star formation is the field of a point source, which can be used when the self-gravity of the gas can be neglected. External fields are also used to describe the gravitational attraction of sink particles, which is discussed in section 3.8.2.

3.6.2 Self-Gravity

To incorporate self-gravitating gas, the Poisson equation (2.38) must be solved numerically for the gravitational potential ϕ at each time step for the present gas density ρ. Assuming two-dimensional geometry and a constant grid spacing $h = \Delta x = \Delta y$ for simplicity, the discretized Poisson equation reads

$$\Delta_h \phi_{i,j} = 4\pi G \rho_{i,j}, \tag{3.12}$$

where the Laplace operator

$$\Delta_h \phi_{i,j} = \frac{\phi_{i+1,j} + \phi_{i-1,j} + \phi_{i,j+1} + \phi_{i,j-1} - 4\phi_{i,j}}{h^2} \tag{3.13}$$

3 Simulations of Star Formation

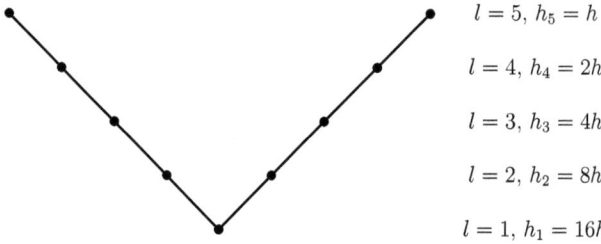

Figure 3.4: Multi-grid V-cycle. The fine grids are restricted to coarser grids with increasing grid spacing h_l. At the coarsest level, the Poisson equation is solved, and afterwards the grid is prolongated to the finer grids. Before restriction and after prolongation, pre-smoothing and post-smoothing iterations are done, respectively.

is discretized via central differences. This is a system of linear equations, which can best be solved using iterative methods (Press et al. 1986). FLASH uses the Gauß-Seidel algorithm with red-black ordering.

Iterative solvers require less memory than direct solvers, but converge slowly because information has to propagate through a large grid. The solution of the system (3.12) can be accelerated if multi-grid methods are employed (Hackbusch 1985). Multi-grid methods introduce a hierarchy of grids with increasingly coarser resolution. The long wavelength components of the solution, which converge the slowest on the fine grid, then correspond to short wavelength components on the coarser grid, which converge fast. In this way, all wavelengths converge simultaneously.

The detailed procedure of one multi-grid cycle is depicted in figure 3.4. It starts at the finest grid, in this case level $l = 5$, with grid spacing $h_5 = h$. Some iterations are done (called pre-smoothing), and then the grid is restricted to level $l = 4$ with the doubled grid spacing $h_4 = 2h$. This is done recursively until the coarsest level $l = 1$ is reached. After the solution is computed iteratively, the grid is prolongated to the next finer level $l = 2$, where again some iterations are done (post-smoothing). This process continues until the finest level $l = 5$ is reached again. Because of the shape of figure 3.4, this procedure is called a V-cycle. The V-cycle is run repeatedly until convergence is achieved.

When adaptive meshes are used, the boundary between patches of different levels of refinement requires special attention. Not only the potential ϕ must be continuous across these boundaries, but also its derivative in order to avoid the introduction of spurious forces. The fine mesh is used to determine these derivatives. The technical details of the FLASH multi-grid implementation can be found in Martin & Cartwright (1996).

Since we want to simulate an isolated region of a molecular cloud, we either have to make our computational domain large enough that we can apply periodic boundary conditions without getting spurious effects, or we choose isolated boundary conditions. The latter case corresponds to a situation with $\rho = 0$ outside the domain, so that $\phi(\boldsymbol{x}) \to 0$ with $|\boldsymbol{x}| \to \infty$. This is achieved following the image mass algorithm of James (1977). The idea is to decompose ϕ as $\phi = \phi_1 - \phi_2$. In a first step, ϕ_1 is calculated from the mass density inside the domain, satisfying a Dirichlet boundary condition $\phi_1 = 0$ at the boundary of the domain ∂D. In the next step, we assume that we can extend the potential ϕ_1 beyond D with $\phi_1(\boldsymbol{x}) = 0$ outside D. This modification can be shown to correspond to an image mass distribution ρ_2 on ∂D. We then solve the Poisson equation for ρ_2 alone, leading to the image potential ϕ_2. Then $\phi = \phi_1 - \phi_2$ satisfies the Poisson equation inside D and the isolated boundary conditions. The obvious drawback of this method is that the multi-grid solver must be run twice, but on the other hand we avoid any boundary effects.

3.7 Radiation

For the modeling of radiation we use the method of hybrid characteristics conceived by Rijkhorst et al. (2006). It is a characteristics-based scheme that neglects the effects of scattering and diffuse radiation, but it includes a detailed treatment of ionization based on the DORIC package developed by Frank & Mellema (1994) and later improved by Mellema & Lundqvist (2002). Rijkhorst et al. (2005) used it to study radiation-driven warped disks and the formation of nebulae. We extended the method significantly by rewriting the raytracing algorithm such that it runs with an arbitrary level of refinement and including the physics of non-ionizing radiation. We describe the method here as it was used in this work.

We illustrate the hybrid characteristics method[1] with the calculation of column densities that are used for the ionizing radiation in section 3.7.1. Here, we assume that the gas consists of atomic hydrogen only, supposing direct dissociation of molecular hydrogen and melting of dust as soon as ionizing radiation becomes dominant. The model of ionization and its effects on the gas is discussed in section 3.7.2. We also consider non-ionizing radiation, for which the optical depth must be calculated separately because the opacity coefficient varies along rays. This calculation is done analogously to the one for the column densities. Non-ionizing radiation is described in section 3.7.3.

3.7.1 Calculation of Column Densities

We use a ray tracing method to calculate column densities for the simulation box. We can conceive two extreme ways how this could be done (see figure 3.5). With the method of long characteristics, a ray is cast from the source to each cell in the domain. The column density at a given point is then calculated along the corresponding ray.

---

[1]We follow the presentation in Rijkhorst et al. (2006), where more details on the implementation can be found.

3 Simulations of Star Formation

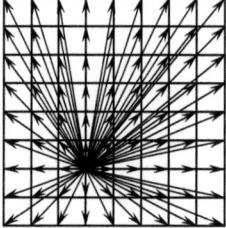 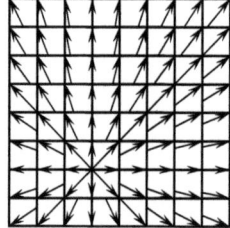

Figure 3.5: Long and short characteristics. The long characteristics method (left) casts a ray from the source to each cell in the domain. This is accurate and parallelizable, but inefficient. The method of short characteristics (right) uses interpolation to find the column densities, which is efficient but necessarily serial.

This method is very accurate and fully parallelizable, but it has the disadvantage that there are a lot of redundant calculations close to the source. The short characteristics method first casts a ray to the cells directly next to the source and then interpolates the column densities to find the values for the cells next to them. This method is computationally very efficient, but the many interpolations introduce rounding errors and, much worse, it is intrinsically serial.

The hybrid characteristics method employs the fact that the computational domain is subdivided into parts and distributed on several processors. Each processor holds a certain patch of the domain. The basic idea is to calculate the column densities in two steps: First, each processor computes the local contribution to the total column densities based on the data it holds using long characteristics. Secondly, long characteristics are used again to find the total column densities from the local contributions by interpolation.

We now explain the algorithm in more detail. For each leaf block, two kinds of local column densities are computed (see figure 3.6). In the first step, long characteristics are cast to the center of each cell in the block. In the second step, rays are cast to the cell corners at the boundary of the block opposing the source for later use as interpolation values.

Now the local contributions must be added. To this end, a ray is cast from each cell to the source, and the traversed leaf blocks are determined. Let $b_{\max}$ be the maximum number of blocks on each processor, b_{L} the local block number and p the processor number. Then

$$b_{\mathrm{G}} = b_{\mathrm{L}} + b_{\max} \cdot p \qquad (3.14)$$

3.7 Radiation

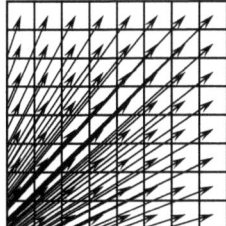 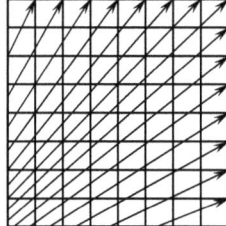

Figure 3.6: Calculation of local column densities. There are two kinds of local column densities. First, long characteristics are cast to each cell in the block (left). Secondly, long characteristics are cast to each cell corner at the block faces opposing the source (right). These values are used later for interpolation.

represents a unique global block number. The problem is now to find the block number given the physical position in the domain. In the original method, a huge array was created, and each element of the array (corresponding to a potential cell on the highest level of refinement) stored the respective global block number b_G. The list of leaf blocks was then created by traversing this block mapping array. Of course, this method is prohibitively memory consuming and against the concepts of AMR. Collapse simulations are impossible using this method.

Instead of creating a large array, we use the fact that FLASH stores hierarchical information about its block structure. Every block in the adaptive-mesh hierarchy has information abouts its parent and child blocks as well as on its neighbors at the same level of refinement. With this information, we can generate the list of traversed leaf blocks by a tree walk along the adaptive mesh hierarchy.

To do this, the bounding box, neighbors, children and parents of each block need to be communicated, which requires no extra memory since this data is available anyway. Starting from a block whose identifier is known, we determine the direction in which the ray leaves the block. If there is a neighboring block on the same level of refinement, it can either still be a leaf block or have one level of children. In the former case the new block is already found, in the latter case the child blocks must be checked. If there is no neighboring block at the same refinement level, the new block must be the neighbor of the parent of the current block in this direction. As this tree walk is effectively a repeated look-up in arrays, it is practically not slower than the original method.

When the list of traversed leaf blocks is known, the stored face values of these blocks can be used for interpolation. Contrary to the discussion in Rijkhorst et al. (2006), the algorithm uses quadratic mean values to interpolate the column densities. Figure 3.7

illustrates the idea. It shows the cell face through which the ray exits the block. This exit point is L_e, and the column density N is known at the cell face corners L_1, L_2, L_3 and L_4. We use quadratic interpolation to first find the column density at L_5 and L_6. To this end, we need to know the length of the ray from the source to each of the points L_i. We do not take the total length of this ray but only the length of the part which lies inside the block to which the face under consideration belongs. Let this local ray length be denoted l_i. Then we determine weights satisfying

$$w_1 l_1^2 + w_2 l_2^2 = l_5^2, \qquad w_1 + w_2 = 1, \qquad (3.15)$$
$$w_3 l_3^2 + w_4 l_4^2 = l_6^2, \qquad w_3 + w_4 = 1. \qquad (3.16)$$

If N_i is the column density at point L_i, then the column densities at L_5 and L_6 are given by

$$N_5 = \sqrt{w_1 N_1^2 + w_2 N_2^2}, \qquad (3.17)$$
$$N_6 = \sqrt{w_3 N_3^2 + w_4 N_4^2}, \qquad (3.18)$$

respectively. For the exit point L_e with local path length l_e, we define weights with respect to L_5 and L_6,

$$w_5 l_5^2 + w_6 l_6^2 = l_e^2, \qquad w_5 + w_6 = 1. \qquad (3.19)$$

The required column density at L_e is then

$$N_e = \sqrt{w_5 N_5^2 + w_6 N_6^2}. \qquad (3.20)$$

Of course, we could have equally well used points L_7 and L_8 instead of L_5 and L_6. The advantage of using the local path lengths to determine the weights is that this procedure introduces only little numerical errors.

3.7.2 Ionizing Radiation

The ionizing radiation does not only change the ionization fraction, but also the temperature of the gas. The collisional ionization rate (2.122) and the radiative recombination rate (2.123) depend on temperature, whereas the photoionization heating rate (2.111) depends on the ionization fraction. To accurately account for these dependencies, we calculate the new ionization fraction and temperature iteratively until they converge. Here, the photoionization heating is counteracted by the metal line cooling curve from Dalgarno & McCray (1972) to prevent overshooting of the temperature. Since the thermal time scale is much smaller than the hydrodynamical time scale, we calculate the new temperature with a sub-cycling scheme.

In each iteration step, the new ionization fraction and temperature is computed. To calculate the ionization fraction, we use the method of Schmidt-Voigt & Köppen (1987). Because

$$x_{\text{HI}} + x_{\text{HII}} = 1 \qquad (3.21)$$

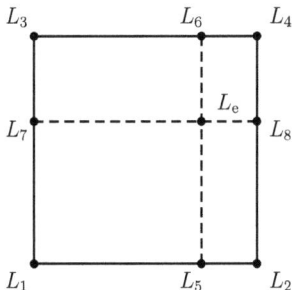

Figure 3.7: Interpolation of column densities. The values at the cell face corner points L_1, L_2, L_3 and L_4 are known. The values at the points L_5 and L_6 are then interpolated from these data according to the length of the corresponding rays. Finally, the column density at the exit point L_e is found using the same interpolation procedure once more.

and the electron density (2.119) is a function of $x_{\text{H\,II}}$, the rate equation (2.118) can be written as a nonlinear ordinary differential equation of $x_{\text{H\,II}}$. But if we treat the electron density as constant, we can analytically solve the resulting linear ordinary differential equation for one time step. The full solution of the nonlinear equation is then found by the iteration process.

To calculate the new temperature, we first define the cooling time step

$$\Delta t_{\text{th}} = \frac{\epsilon^i}{\left|\Gamma_{\text{ph}}^i - \Lambda_{\text{ml}}^i\right|}, \qquad (3.22)$$

where ϵ is the internal energy density, Γ_{ph} is the photoionization heating rate and Λ_{ml} is the metal line cooling rate. The photoionization heating rate Γ_{ph} is constant over one sub-cycling process, but the metal line cooling rate Λ_{ml} depends on temperature and thus changes after each step. The new internal energy density is then given by

$$\epsilon^{i+1} = \epsilon^i + (\Gamma_{\text{ph}}^i - \Lambda_{\text{ml}}^i)\Delta t_{\text{th}}. \qquad (3.23)$$

This is repeated until the thermal time steps sum up to a hydrodynamical time step.

3.7.3 Non-ionizing Radiation

The optical depths for the non-ionizing radiation can be obtained totally analogously to section 3.7.1. The only difference here is that the opacity also depends on temperature, which causes a stronger spatial variation of the optical depth than in the case of ionizing radiation. However, this does not introduce any numerical complications.

3 Simulations of Star Formation

The heating of the gas by the non-ionizing radiation is much smaller than by the ionizing radiation, so that it can be considered seperately. This heating is done together with the dust cooling, which is described in section 3.5.

3.8 Sink Particles

Fully resolving gravitational collapse would strictly require a resolution down to the size of a protostar, which is several solar radii (Stahler & Palla 2004). Numerically, such a high resolution is prohibitive if the large-scale dynamics is of interest because it is the smallest grid size which determines the time step. Instead, the idea is to represent gravitationally unstable regions beyond the grid resolution with sink particles (Bate et al. 1995, Krumholz et al. 2004).

3.8.1 Creation and Accretion

In the simplest implementation, sink particles can be dynamically created by a density threshold criterion. Let Δx be the grid size at the highest refinement level. If we want to resolve the local Jeans length with at least n cells, then

$$\lambda_J \geq n\Delta x \tag{3.24}$$

must hold. As discussed in section 3.2.2, $n \geq 4$ should be taken in any case. Using equation (2.47), this translates into the condition

$$\rho_0 \leq \frac{\pi c_s^2}{G(n\Delta x)^2}. \tag{3.25}$$

By assuming a low temperature, we can estimate the smallest density $\rho_{\max}$ above which this inequality is not satisfied anymore. If a cell above this threshold density is found, a sink particle is created and all mass $\rho > \rho_{\max}$ within a certain accretion radius r_{sink} is accreted onto the sink particle[2]. We take this sink particle radius to be

$$r_{\text{sink}} = \sqrt{\frac{\pi c_s^2}{G\rho_{\max}}}. \tag{3.26}$$

The creation of sink particles can be made more sophisticated by applying additional checks. This may be necessary because gas above the threshold density does not need to be bound. These additional criteria check the existence of a minimum in the gravitational potential, convergence of the flow, and the relation of thermal and kinetic energy to the gravitational energy of the gas (Bate et al. 1995, Federrath et al. 2009). This avoids the spurious creation of sink particles in shocks and creation of multiple sink particles in large overdense regions.

---

[2] To avoid confusion with the accretion radius r_{acc} in the accretion heating model, we call this radius the sink radius r_{sink}.

3.8 Sink Particles

The accretion of overdense regions by sink particles is a two-step procedure. In the first step, every cell is checked to satisfy $\rho(\boldsymbol{x}) \leq \rho_{\max}$. If this condition is violated, a loop over all sink particle positions $\boldsymbol{y}$ tries to find a corresponding particle with $|\boldsymbol{x} - \boldsymbol{y}| \leq 2r_{\text{sink}}$. The factor of 2 assures that a newly formed sink particle has no overlap with already existing ones. If such a particle is not found, the additional checks are applied, and if all conditions are satisfied, a new sink particle is created. At this point of the algorithm for each overdensity which should be accreted onto a sink particle there is a corresponding sink particle available. In the second step of the procedure, we again loop over all blocks with an overdensity. If a corresponding sink particle exists and the gas in this cell converges towards this sink particle, the overdense gas is accreted.

The accretion process not only changes the mass of the sink particle, but also its linear momentum. Because the gas taken out of the domain carries momentum, we add this momentum to the sink particle to ensure momentum conservation. If one thinks about the sink particles as representing protostars, this is of course just a crude approximation, since the material would really spiral in and be accreted via an accretion disk rather than just falling onto the protostar with the same momentum it had several 100 or even 1000 AU away. In practice this is of little relevance since the dynamics of the sink particles is dominated by the gravitational interaction with each other and the gas.

3.8.2 Gravitational Field

The mass hidden in the sink particles can still act gravitationally with the gas. This is done by adding the gravitational acceleration by the sink particle $\boldsymbol{g}_{\text{sp}}$ to the acceleration by the gas $\boldsymbol{g}_{\text{gas}}$, which is found by solving the Poisson equation (see section 3.6.2). The total gravitational acceleration

$$\boldsymbol{g}_{\text{tot}} = \boldsymbol{g}_{\text{gas}} + \sum_{i=1}^{N_{\text{sp}}} \boldsymbol{g}_{\text{sp},i} \qquad (3.27)$$

is the sum of the gas acceleration and the contribution of all N_{sp} sink particles. The total acceleration $\boldsymbol{g}_{\text{tot}}$ is then used as a source term in the Euler equations.

Since the gravitational acceleration of a point mass diverges at its position, $\boldsymbol{g}_{\text{sp}}$ must be softened at a finite radius r_{soft}. This softening radius should be chosen such that $r_{\text{soft}} \leq r_{\text{sink}}$. We define $\boldsymbol{g}_{\text{sp}}$ such that it is identical to the acceleration of a point source for $r \geq r_{\text{soft}}$ and decreases linearly with $r \to 0$. This leads to the prescription

$$\boldsymbol{g}_{\text{sp}}(\boldsymbol{r}) = \begin{cases} -G\dfrac{m}{r_{\text{soft}}^3}\boldsymbol{r} & r \leq r_{\text{soft}}, \\ -G\dfrac{m}{r^3}\boldsymbol{r} & r > r_{\text{soft}} \end{cases} \qquad (3.28)$$

with the mass of the sink particle m.

61

3.8.3 Equation of Motion

The only forces that have to be accounted for when solving the equation of motion of the sink particles are gravitational since sink particles do not couple to the flow and the change of momentum due to accretion is set when the gas is accreted. Hence, the sink particles are only subject to the gravitational acceleration $\boldsymbol{a} = \boldsymbol{g}_{\text{tot}}$ from equation (3.27).

The sink particles are advanced by a second-order leapfrog method. It uses time-centered velocities and stored accelerations to keep the method of second order in time with a variable time step. Let Δt^n be the present and Δt^{n-1} the old time step, then we define the coefficients

$$C_n = \frac{1}{2}\Delta t^n + \frac{1}{3}\Delta t^{n-1} + \frac{1}{6}\left(\frac{(\Delta t^n)^2}{\Delta t^{n-1}}\right) \tag{3.29}$$

and

$$D_n = \frac{1}{6}\left(\Delta t^{n-1} - \frac{(\Delta t^n)^2}{\Delta t^{n-1}}\right). \tag{3.30}$$

Given the present and old acceleration $\boldsymbol{a}^n$ and $\boldsymbol{a}^{n-1}$, respectively, as well as the old velocity $\boldsymbol{v}^{n-1/2}$, we can calculate the new velocity as

$$\boldsymbol{v}^{n+1/2} = \boldsymbol{v}^{n-1/2} + C_n \boldsymbol{a}^n + D_n \boldsymbol{a}^{n-1}. \tag{3.31}$$

The old position $\boldsymbol{x}^n$ is updated via

$$\boldsymbol{x}^{n+1} = \boldsymbol{x}^n + \boldsymbol{v}^{n+1/2}\Delta t^n. \tag{3.32}$$

4 Results

> The purpose of computing is insight, not numbers.
>
> (Richard Hamming,
> Numerical Methods for Scientists and Engineers)

As a first step towards numerical simulations of star formation with ionization feedback, we have tested the performance of the ionization module and compared the results with the analytical Spitzer solution (section 4.1). The agreement is very good. Next, we have applied the new method to study the situation of "cloud-crushing", the interaction of an ionization front with a dense clump of molecular gas. Here, we put a special focus on the generation of turbulence in the wake behind the ionized clump (section 4.2). The heart of this work are the collapse simulations presented in section 4.3. Collapse simulations of this kind, which include the dynamical formation of an accretion disk and the build-up and growth of an H II region, have never been conducted so far. They lead to a novel understanding of the interaction between ionizing radiation and the accretion flow around massive protostars and connect as diverse areas as H II region morphologies, time variability of UC H II regions, ionization-driven bipolar outflows and the upper mass limit.

4.1 Verification

Every numerical code must be verified before it can be used. The obligatory test against analytically solvable problems, experimental findings or other numerical codes with known reliability is crucial to get confidence in results that are totally new. The FLASH code itself has undergone a large amount of testing, for example in the context of hydrodynamical instabilities (Calder et al. 2002). Rijkhorst et al. (2006) verified the capability of the raytracing module of calculating column densities and casting shadows accurately. Iliev et al. (2006) performed the first dynamical tests of the hybrid characterstics method in a cosmological setting. They found that it has some problems with tracking very fast R-type ionization fronts. This raises the question whether we can adequately model the D-type ionization fronts which are relevant for massive star formation. The expansion velocity of these D-type ionization fronts is determined by the sound speed of the ionized gas (see section 2.4.6), which leads to a much slower expansion compared to R-type fronts, which do not dynamically follow the gas. This suggests that D-type ionization fronts may still be handled sufficiently well despite the problems with R-type fronts. To check this, we simulate the expansion of a D-type ionization front around an O-star in a homogeneous medium and compare

4 Results

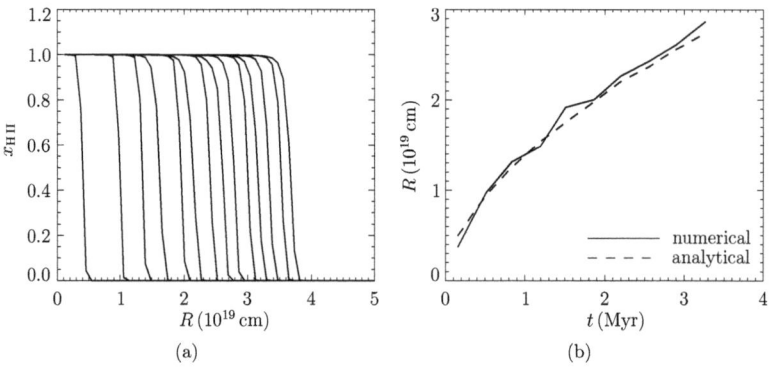

Figure 4.1: Expansion of a D-type ionization front in a homogeneous medium. (a) Spherically averaged ionization fraction $x_{\text{H\,II}}$ as function of radius R at different times; time increases from left to right. The time between successive curves is roughly 0.35 Myr. At the ionization front, the ionization fraction drops abruptly from 1 to 0. (b) Radius R of the ionization front as function of time t. The analytical data is dashed, the numerical data is solid. The wiggling of the numerical curve is caused by waves which deform the ionization front slightly. Despite these irregularities, the curves agree very well.

the results with the analytical Spitzer solution (2.135).

This analytical solution has only two parameters, the Strömgren radius R_S and the sound speed c_s in the ionized gas. The Strömgren radius depends, by definition, on the initial density and, via the recombination coefficient (2.123), on the temperature in the ionized gas. Since we do not resolve the Strömgren radius in this test simulation, we pick a typical temperature of 10^4 K. The initial number density in the setup is $n_\text{H} = 2000\,\text{cm}^{-3}$. The sound speed must be taken inside the expanding H II region, but because of gas-dynamical processes like waves running away from the source, the temperature of the ionized gas turns out to be inhomogeneous and time-dependent, so we take the volume average in the H II region.

To compare the prediction based on the analytial solution with the actual numerical data, we need to determine the radius of the H II region in the simulation. To this end, we calculate the spherical average of the ionization fraction $x_{\text{H\,II}}$ and find the radius where it decreases from 1.00 to 0.97. Figure 4.1(a) shows this spherical average as a function of radius for different snapshots of the simulation. The plot demonstrates that the ionization fraction drops sharply from a high to a low value, so the inferred radius does not depend much on the exact threshold.

The analytical and numerical data for the H II region radius is plotted in figure 4.1(b). Given the simplified assumptions of the analytical solution, they show a remarkably good agreement. The wiggling of the numerical data is due to waves running through the H II region which deform the shell and add additional dynamics to the problem. This is not accounted for in the simple Spitzer solution. Hence, within the typical numerical uncertainties, the expansion velocities of D-type ionization fronts are confirmed to be correct.

4.2 Driving of Turbulence by Ionization Fronts

As a first application of the ionization module, we study the properties of turbulence in the wake of an ionization front that hits a Bonnor-Ebert sphere. To generate the ionization front, we place a source at the corner of the computational domain. The surrounding gas gets photoionized, a D-type ionization front expands and finally hits the Bonnor-Ebert sphere. Though the Bonnort-Ebert sphere gets compressed by the shock, star formation is not triggered because the photoevaporation timescale is shorter than the collapse timescale in the shock-compressed gas. As the shock front runs over the Bonnor-Ebert sphere and interacts with its dense material, it creates a turbulent wake behind the sphere. We focus on the generation and evolution of turbulence in this wake.

The computational domain has dimensions $(6.0 \times 2.4 \times 2.4) \cdot 10^{19}$ cm with an effective resolution of $512 \times 256 \times 256$ cells. We place a self-gravitating supercritical Bonnor-Ebert sphere with mass $M = 100\,M_\odot$ in the center of the domain. It slowly rotates with an angular velocity of $\Omega = 7.83 \cdot 10^{-15}$ rad/s around the z-axis corresponding to a ratio of rotational to gravitational energy of $\beta = 0.02$. Additionally to the rotational velocity, a turbulent velocity field with a magnitude of at most 50 % of the sound speed is added. The temperature of the Bonnor-Ebert sphere is $T = 20\,\text{K}$, while the ambient gas has $T = 90\,\text{K}$.

The ionizing source is located at the left hand side of the computational domain at $(0.0, 1.2, 1.2) \cdot 10^{19}$ cm. It has a temperature of $T_{\text{star}} = 29,200\,\text{K}$ and a luminosity of $L = 24,000\,L_\odot$, representing a B0 star. The radiation heats the interstellar gas to $T \approx 6 \cdot 10^4\,\text{K}$.

When the simulation starts, the gas next to the source becomes ionized and heated. An ionization front accompanied by a shock expands into the ISM. When the shock hits the clump, the clump material is swept away and thereby compressed heavily. The shock leaves behind a fully turbulent gas. The hot ionized gas is being mixed with the cold clump material while the radiation heats it up continuously. After the shock front passes the clump, the former clump disperses completely.

The time sequence of this process is as following. The shock touches the outer edge of the clump at $t = 0.40\,\text{Myr}$. At $t = 0.53\,\text{Myr}$, it reaches the clump center at $x = 3 \cdot 10^{19}$ cm, and at $t = 0.76\,\text{Myr}$, the front has totally enclosed the clump. Of course, the dense material delays the shock, so that the wings of the front can propagate faster. They enter the shadow zone and meet in the middle at $t = 0.77\,\text{Myr}$. The clash of these wings is an important driving mechanism of the turbulence seen in these simulations. It builds up an extended turbulent wake while the ionization front

4 Results

propagates further into the ISM. At $t = 1.09\,\mathrm{Myr}$, the shock reaches the boundary of the computational domain at $x = 6 \cdot 10^{19}\,\mathrm{cm}$. The simulation stops at $t = 1.49\,\mathrm{Myr}$ when the cloud has dissolved. Some of the stages of this time evolution are depicted in figure 4.2.

We start our analysis of the turbulence properties with a brief look at the energy balance in the flow (see figure 4.3(a)). The plot shows the total energy E, internal energy E_{int} and kinetic energy E_{kin} in erg as a function of time t. The B star provides energy to the system by ionizing material. This contribution is predominantly transferred into internal energy by the photoionization heating, only a small fraction is converted into kinetic energy. For example, at $t = 0.5\,\mathrm{Myr}$, the ratio of internal over kinetic energy is $E_{\mathrm{int}}/E_{\mathrm{kin}} \approx 57$.

Since the luminosity of the star is constant in time, the energy transferred from the star to the gas in the computational domain grows linearly. This explains qualitatively the form of $E(t)$ in figure 4.3(a). A quantitative analysis is difficult however, since geometric effects have to be taken into account appropriately. One would have to account for the fact that the star emits its radiation isotropically, while it is not at the center of a spherically symmetric computational domain, but on one face of a Cartesian box.

The simulation shows that the cloud-crushing flow is largely dominated by supersonic motion. This is because the cloud material is cold, so that the sound speed is much lower than in the hot gas behind the ionization front, where a wind with $\mathcal{M} \approx 0.2\text{--}0.4$ is observed. The cause of the wind is that the photoionization heating is stronger close the source, which leads to a pressure gradient and a corresponding flow. Since the wind prevails in the largest part of the domain, namely the hot postshock gas, the mean Mach number $\mathcal{M}_{\mathrm{mean}}$ is always below unity, while the maximum Mach number $\mathcal{M}_{\mathrm{max}}$, which is reached at crushing, can be greater than 6.

Figure 4.3(b) depicts $\mathcal{M}_{\mathrm{max}}$ and $\mathcal{M}_{\mathrm{mean}}$ as a function of time. The maximum Mach number traces the shock ahead of the ionization front. Within a time of $0.15\,\mathrm{Myr}$, the shock accelerates up to a constant velocity of $\mathcal{M} = 4$. For a short time of another $0.15\,\mathrm{Myr}$ the velocity seems to saturate, but the continuous photoionization heating accelerates the shock again up to $\mathcal{M} = 6.5$. At this point, when $\mathcal{M}_{\mathrm{max}}$ is maximal, the shock collides with the dense clump. As it hits the high-density gas, the shock front moves more slowly. The gas decelerates, so that $\mathcal{M}_{\mathrm{max}}$ decreases again. However, the wings of the shock that were not affected by the clump can enter the shadow zone, which leads to a peak in $\mathcal{M}_{\mathrm{max}}$ after $0.7\,\mathrm{Myr}$. Then these wings collide, which stops the motion in y-direction, so that the Mach number decreases further. But the collision of the wings also leads to an acceleration in positive x-direction, which can be seen in a series of peaks from $0.8\,\mathrm{Myr}$ to $1.1\,\mathrm{Myr}$. After $1.1\,\mathrm{Myr}$, the shock leaves the computational domain, resulting in a sharp drop in $\mathcal{M}_{\mathrm{max}}$. This demonstrates that the highest Mach numbers are only reached in the shock front itself, not in the shock-generated turbulence behind the front. The motion in the turbulent wake is mostly supersonic with $\mathcal{M}$ below 3. The mean Mach number grows continuously until the shock leaves the domain, whereafter $\mathcal{M}_{\mathrm{mean}}$ declines slowly. The heating of the gas does not change $\mathcal{M}_{\mathrm{mean}}$.

4.2 Driving of Turbulence by Ionization Fronts

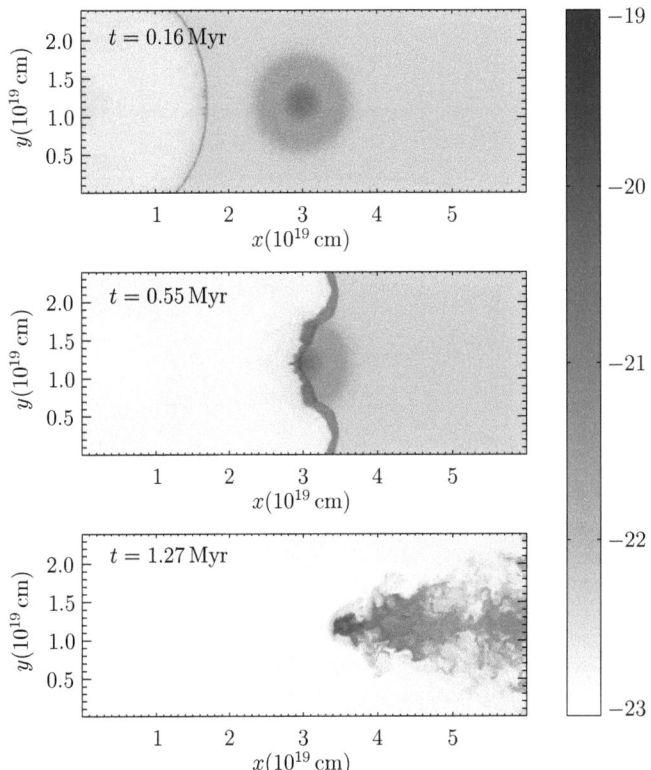

Figure 4.2: Main stages of clump evaporation. The figure shows two-dimensional cuts of the mass density $\log \rho$ in $\mathrm{g\,cm^{-3}}$ in the midplane. The different snapshots are taken at times $t = 0.16\,\mathrm{Myr}$, $t = 0.55\,\mathrm{Myr}$ and $t = 1.27\,\mathrm{Myr}$, respectively. The turbulent wake behind the former clump at the last stage is nicely visible.

4 Results

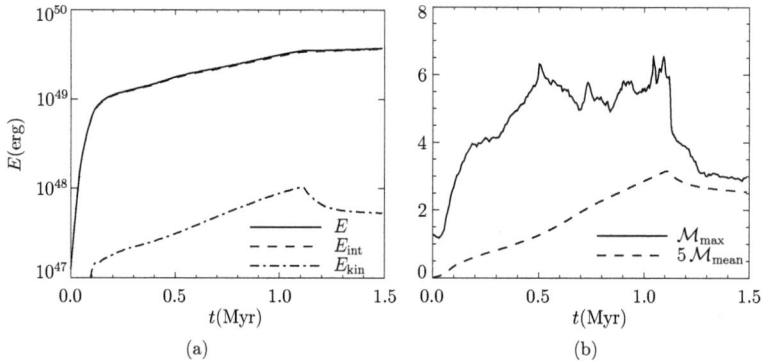

Figure 4.3: Time evolution of energies and Mach numbers during cloud-crushing. (a) The total energy E, internal energy E_{int} and kinetic energy E_{kin} are shown as function of time t. The major contribution to E comes evidently from E_{int}, while E_{kin} is between one and two orders of magnitude smaller. The point in time where the shock, which carries most of the kinetic energy, leaves the computational domain can be distinguished easily. (b) The average and maximum Mach numbers in the flow provide information on the dominance of shocks in the flow. To compare these two numbers, the mean Mach number $\mathcal{M}_{\text{mean}}$ is displayed amplified by a factor of 5. The peaks in $\mathcal{M}_{\text{max}}$ can be associated with events in the cloud-crushing scenario.

4.2 Driving of Turbulence by Ionization Fronts

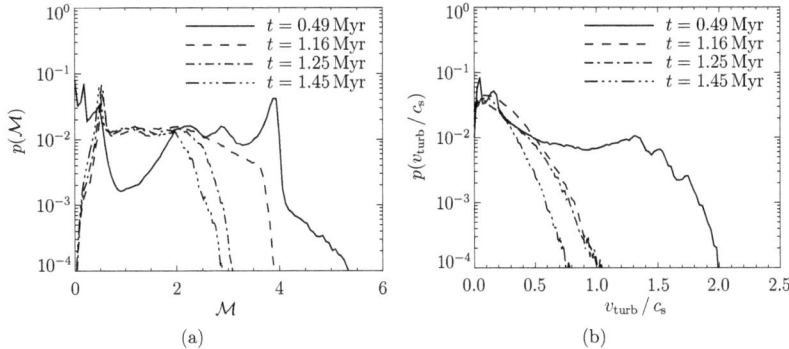

Figure 4.4: Mach number PDFs during cloud-crushing. (a) The mass-weighted PDFs of the Mach number $\mathcal{M}$ between times $t = 0.49\,\text{Myr}$ and $t = 1.45\,\text{Myr}$. The shock front shows up as a peak in the regime of high Mach numbers for the first point in time. At $t = 1.16\,\text{Myr}$, the shock has left the computational domain, and the turbulence with its supersonic Mach numbers decays. (b) The mass-weighted PDFs of the "turbulent Mach number" v_turb/c_s with $v_\text{turb} = \sqrt{v_y^2 + v_z^2}$ for the same points in time as in figure (a). At cloud-crushing, the turbulent field v_turb is slightly supersonic, while the flow in the wake is only subsonic. This shows that the bulk motion contributes significantly to the high total Mach numbers.

The different stages of the simulation can also be recognized in the probability density functions (PDFs) of the Mach number $\mathcal{M}$. Figure 4.4(a) shows mass-weighted PDFs at the moment of cloud-crushing ($t = 0.49\,\text{Myr}$) and after the shock has left the domain ($t = 1.16\,\text{Myr}$ to $t = 1.45\,\text{Myr}$). At the crushing time, the high $\mathcal{M}$ above 2 all belong to the shock. The shock then excites supersonic turbulence in the wake, but away from the shock $\mathcal{M}$ above 3 is very rare. While most of the dense gas is supersonic, most of the domain is dominated by low Mach number flows, both at crushing time and afterwards.

The PDFs of the total Mach number $\mathcal{M}$ should be compared with the Mach number given only by the turbulent velocity fluctuations, $v_\text{turb} = \sqrt{v_y^2 + v_z^2}$, which is v_turb/c_s, where c_s is the local speed of sound. These plots are shown in figure 4.4(b). Since the bulk motion in x-direction is no longer taken into account, the Mach numbers are significantly lower. At the moment of cloud-crushing, the turbulent Mach number is only slightly supersonic, while it is totally subsonic afterwards. Hence, the bulk motion of the shock (and also the transport of momentum by the wind) is important to reach the high Mach numbers observed above.

Another quantity of interest is the fraction of mass which moves supersonically. In figure 4.5 we plot the ratio of the mass of supersonic gas M_sup and the total gas mass in the computational domain M_tot. Despite of the complicated mixing processes,

4 Results

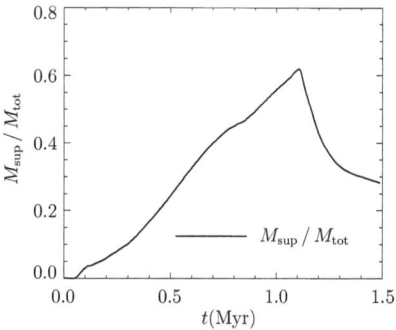

Figure 4.5: Fraction of supersonic gas during cloud-crushing. The fraction of gas in supersonic motion $M_{\text{sup}}/M_{\text{tot}}$ as a function of time t. Although most of the energy input goes into internal energy, more than half of the gas moves supersonically when the shock leaves the domain.

$M_{\text{sup}}/M_{\text{tot}}$ grows roughly linearly until the shock leaves the domain. It is surprising that a significant part of the gas moves supersonically, altougth most of the energy input is converted into internal energy (see figure 4.3(a)). When the shock front leaves the domain, more than 60 % of the gas in our domain is in supersonic motion.

In figure 4.6, we enlarge a part of figure 4.2 belonging to the snapshot at $t = 1.27\,\text{Myr}$ and show additionally to the mass density also the temperature and the velocity components v_x and v_y. The remains of the dense core have a temperature around $T \approx 300\,\text{K}$, while the ambient ionized medium is at $T \approx 3.5 \cdot 10^4\,\text{K}$. The high temperatures in the environment give rise to the "rocket effect", which accelerates the gas in positive x-direction (Oort & Spitzer 1955). This is because the cold gas at the surface of the clump facing the star becomes heated. Thus, it expands into the postshock medium, carrying momentum with it, and consequently the clump accelerates.

The cloud-crushing leads to vortical structures in the wake as can be seen from the lower plots in figure 4.6. The two largest structures around $x \approx 4.2 \cdot 10^{19}\,\text{cm}$ and $y \approx 0.7 \cdot 10^{19}\,\text{cm}$ or $y \approx 1.7 \cdot 10^{19}\,\text{cm}$, respectively, even have slightly negative v_x and so does a region around the tip of the former core at $x = 3.3 \cdot 10^{19}\,\text{cm}$ and $y = 1.2 \cdot 10^{19}\,\text{cm}$. Averaged over the whole volume, however, the wind, wich moves with $v_x \approx 1.5 \cdot 10^6\,\text{cm/s}$, causes a bulk motion of the gas in positive x-direction. In order to measure the turbulent components of the velocity field, it is better to focus on the transversal directions. The peak amplitude of the velocity component in y-direction, for example, is about half of the maximum velocity in x-direction. The large vortical structures discussed above have total velocities around $10^6\,\text{cm/s}$. The upper vortex rotates clockwise, while the lower vortex rotates counter-clockwise.

A connection to observations of molecular clouds can be made by looking at velocity profiles along a certain line-of-sight (Ossenkopf & Mac Low 2002). Examplary, we show in figure 4.7 a profile for v_y of the dense core after crushing at time $t = 1.27\,\text{Myr}$

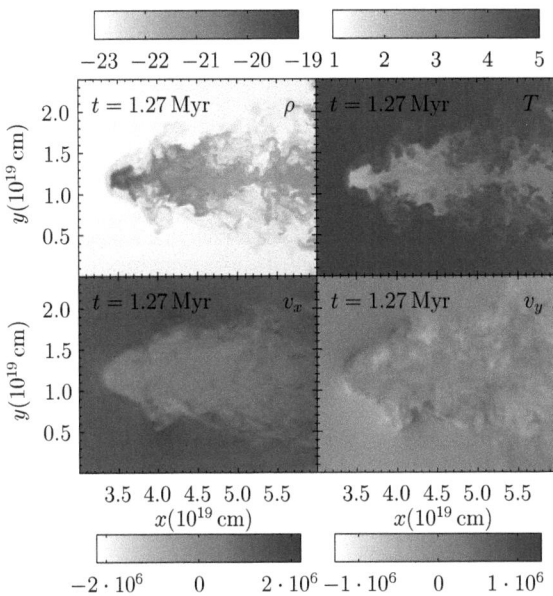

Figure 4.6: Slices through the turbulent clump. The midplane cuts at $t = 1.27\,\mathrm{Myr}$ of the density $\log \rho$ in $\mathrm{g\,cm^{-3}}$, the temperature $\log T$ in K, the velocity in x-direction v_x in $\mathrm{cm\,s^{-1}}$ and the velocity in y-direction v_y in $\mathrm{cm\,s^{-1}}$ are compared with each other. The remains of the dense clump are still cold, while the material in the wake is already heating up. While v_y is a measure for the turbulent velocity fluctuations, v_x shows the bulk motion of the gas.

4 Results

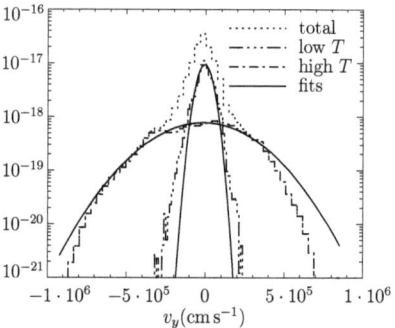

Figure 4.7: Line-of-sight mass-weighted histograms of the turbulent clump. Figure shows histograms for v_y around the dense core after cloud-crushing at time $t = 1.27\,\text{Myr}$ (in arbitrary units). The figure shows both the total data and the low temperature $T \leq 10^3\,\text{K}$ and high temperature $T \geq 10^4\,\text{K}$ cuts. Clearly, the high velocity contributions stem from the high T gas, whereas the peak around zero relates mainly to the low T gas and a warm envelope with T between the two cuts. Also shown are Gaussian fits to the low and high T cuts that give insight into the strength of the turbulent velocity fluctuations and can be compared with observed spectral line widths.

(compare also figure 4.6). The width of the beam is $3.0 \cdot 10^{19}\,\text{cm} \leq x \leq 4.0 \cdot 10^{19}\,\text{cm}$ and $0.7 \cdot 10^{19}\,\text{cm} \leq z \leq 1.7 \cdot 10^{19}\,\text{cm}$, while y ranges over the whole box width. Note that the profile is a mass-weighted histogram, not a normalized PDF. We do the same calculation again, but this time only considering gas that has $T \leq 10^3\,\text{K}$ (low T case) or $T \geq 10^4\,\text{K}$ (high T case). From the figure we see that the high velocity tails are exclusively related to the high temperature gas. The peak at low velocities comes mainly from both the low temperature medium as well as a warm envelope with temperatures between the cuts $10^3\,\text{K} \leq T \leq 10^4\,\text{K}$. Additionally to the histograms, we have fitted Gaussians to the low and high T data. Their variance gives a measure for the turbulent velocity and can be related to the width of spectral lines that are being oberserved.

To summarize our findings, we have seen that cloud-crushing by ionization fronts can lead to short-living supersonic turbulence. Altough only a minute fraction of the energy input is converted into kinetic energy, up to 60 % of the affected gas is supersonic. While it is mainly the cold gas that is highly supersonic, it is the hot gas that moves the fastest. The bulk motion of the shock is an important contribution to the supersonic flow, since the transversal fluctuations are at best slightly supersonic.

Although ionizing radiation injects a significant amount of energy into the ISM, it does not seem to be an important driving mechanism of interstellar turbulence on a global scale (Mac Low & Klessen 2004). However, the cloud-crushing process generates a considerable amount of turbulence locally in the wake of the cloud. We find that the motion of the cloud material is mostly supersonic while the ambient gas behind the front moves only subsonically. The continuous heating limits the lifetime of the dense material, but the supersonic motions are maintained until the cloud disperses. This is contrary to the situation in jet-clump interactions, where the situation is less clear with some studies showing mostly subsonic motions (Banerjee et al. 2007) while others claim supersonic velocity fields (Nakamura & Li 2007). Although the clump is compressed considerably, the photoionization front evaporates the clump too quickly that star formation cloud be triggered.

4.3 Collapse Simulations

We have performed several collapse simulations to study the effects of ionization feedback in massive star formation. These are the first three-dimensional collapse simulations that include heating by both ionizing and non-ionizing radiation. We employ the novel numerical methods laid out in chapter 3. We follow the gravitational collapse until we reach the maximum level of refinement. If the gas is still collapsing further and we cannot resolve the Jeans length anymore, we dynamically form a sink particle. The sink particle is coupled to the raytracing module via a prestellar model. It can emit both ionizing and non-ionizing radiation; the latter is dominated by the accretion heating. The radiative transfer method can handle an arbitrarily large number of sources. For the first time, numerical simulations of star cluster formation with radiative feedback by ionizing and non-ionizing radiation are possible.

We start our calculations with a molecular cloud with a mass of $1000 M_\odot$ and an initial temperature of 30 K. The cloud has a flat inner region with 0.5 pc radius, surrounded by a region in which the density drops as $r^{-3/2}$. It is in solid-body rotation with a ratio of rotational to gravitational energy $\beta = 0.05$.

This setup was chosen initially to probe the upper mass limit. But it turns out that the disk which develops during the course of the simulation is gravitationally unstable and fragments, giving rise to multiple star formation. We have artificially suppressed multiple sink formation with the Jeans heating to address the question whether ionizing radiation can stop protostellar accretion (section 4.3.1). The time evolution of the protostar dramatically changes when a whole star cluster is allowed to form (section 4.3.2). Both simulations show very similar disk fragmentation (section 4.3.3), bipolar outflows (section 4.3.4), H II region morphologies (section 4.3.5) and time variability (section 4.3.6).

An overview of the collapse simulations is given in table 4.1. The high resolution simulations have a cell size at highest refinement level of 98 AU, whereas the low resolution simulations have a cell size of 196 AU at highest level of refinement, so they differ only by a factor of two. For the high resolution simulations (Run A, Run B and Run D), we use $\rho_{\mathrm{max}} = 7 \cdot 10^{-16}\,\mathrm{g\,cm^{-3}}$ and $r_{\mathrm{sink}} = 590\,\mathrm{AU}$, for the lower resolution simulations (Run Ca and Run Cb), we take $\rho_{\mathrm{max}} = 1 \cdot 10^{-16}\,\mathrm{g\,cm^{-3}}$ and $r_{\mathrm{sink}} = 1319\,\mathrm{AU}$. The three simulations with a single sink particle (Run A, Run Ca and Run Cb) use the Jeans heating to avoid runaway overdensities. The simulation without feedback

4 Results

Name	Resolution	Multiple Sinks	Radiative Feedback
Run A	high	no	yes
Run B	high	yes	yes
Run Ca	low	no	yes
Run Cb	low	no	yes
Run D	high	yes	no

Table 4.1: Overview of collapse simulations. We have run high and low resolution simulations with a single sink particle to probe the upper mass limit as well as high resolution simulations with multiple sink particles to study stellar cluster formation with and without radiative feedback.

(Run D) has neither ionizing nor non-ionizing radiation feedback. It is used as a control run to compare with Run B. The SFE at the moment when the simulations are stopped is between 7 and 15 %, but all simulations show ongoing star formation and accretion at the end.

4.3.1 Upper Mass Limit

As a first approach to the question whether ionizing radiation may stop the accretion process, we have conducted collapse simulations at lower resolution. This allows us to run the simulation with a larger time step and thus to make faster progress. To make sure that the accretion histories at low resolution are consistent with the higher resolution simulations, we have also conducted a high resolution simulation, but we have not run it up to the same protostellar mass as soon as it became clear that the higher resolution simulation would not yield different results than the lower resolution runs. One of the lower resolution simulations, Run Ca, starts with identical initial conditions as the high resolution Run A, while Run Cb has an additional $m = 2$-perturbation of 10 % of the gas density (Boss & Bodenheimer 1979). This allows us to check that the results do not depend on details of the initial conditions. In order to be sure that enough gas is available to the massive protostar, we suppress secondary sink formation with the Jeans heating in all these runs.

The accretion history for these simulations is depicted in figure 4.8. One can clearly see that the additional perturbation as well as the change in resolution has no influence on the overall accretion behavior. Run A accretes at a mean accretion rate of $5.9 \cdot 10^{-4} \, M_\odot \, \text{yr}^{-1}$, while Run Ca and Run Cb accrete at mean rates of $4.6 \cdot 10^{-4} \, M_\odot \, \text{yr}^{-1}$ and $5.8 \cdot 10^{-4} \, M_\odot \, \text{yr}^{-1}$, respectively. Run Ca was stopped when the massive protostar reached $100 M_\odot$, Run Cb was stopped at $94 M_\odot$. These very high masses constitute the end of the available ZAMS data that determines the strength of the ionizing radiation. At no point of the simulations does the accretion rate of any run drop below $10^{-5} \, M_\odot \, \text{yr}^{-1}$. This suggests that ionizing radiation is unable to stop protostellar accretion.

Although the ionizing radiation blows away substantial parts of the accretion disk, it can never create a bubble around the protostar that could reverse the accretion

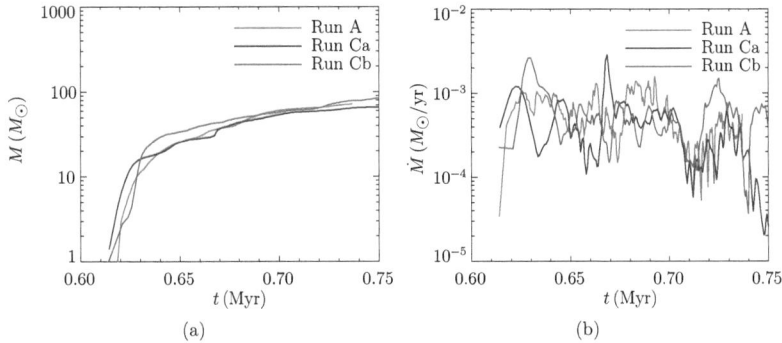

Figure 4.8: Accretion history of single sink simulations. (a) Protostellar masses. The accretion process proceeds very similar in all three simulations. The high resolution simulation Run A was stopped at $72 M_\odot$, the lower resolution simulations have run up to $100 M_\odot$ (Run Ca) and $94 M_\odot$ (Run Cb). (b) Accretion rates. The mean accretion rates of $5.9 \cdot 10^{-4}\,\mathrm{M_\odot\,yr^{-1}}$, $4.6 \cdot 10^{-4}\,\mathrm{M_\odot\,yr^{-1}}$ and $5.8 \cdot 10^{-4}\,\mathrm{M_\odot\,yr^{-1}}$, respectively, are very similar. This suggests that the accretion behavior depends neither on the initial conditions nor on the numerical resolution. There is no indication that ionization feedback may ever stop accretion.

flow. It is worth noting that this is still the case when the protostar is moving radially outwards into lower density gas. On the contrary, the sink particle in Run A is in the lowest density material by the end of the simulation, but it only has a tiny H II region because it is quenched by the strong accretion flow. Dense filaments in the disk shield the ionizing radiation and do not allow the H II region to expand symmetrically around the protostar. The chaotic interaction of the ionizing radiation with the accretion flow causes the H II region to flicker and drives bipolar outflows perpendicular to the disk. These phenomena are discussesed in more detail in the following sections.

Although the results demonstrate that ionizing radiation does not cut off accretion when a large enough gas reservoir is available to accrete from, the star formation scenario in the single sink simulations is not particularly realistic. The fragments which develop in the disk would form additional protostars if we would not artificially heat them up. This is in agreement with the observation that massive stars form in clusters. However, multiple sink formation allows the gas reservoir to be split between several protostars. Since the luminosity scales with a power of mass greater than unity, this means that the total ionizing luminosity is smaller in the multiple sink simulation. On the other hand, the gas reservoir is now used up by several protostars, which weakens the accretion flow on each individual star. As we show in the following section, the second effect is dominant, so that ionizing radiation can create a bubble when a star cluster is allowed to form.

4.3.2 Star Cluster Formation

Our results from the previous section suggest that the accretion flow onto massive stars cannot be stopped by ionizing radiation. Thus, one might expect that equally massive stars could form in the stellar cluster as in the single sink runs, but this does not happen. Instead, the mass growth of individual objects is halted by gravitational fragmentation of the disk and the subsequent competition of massive stars with lower-mass companions for the common gas reservoir of the disk. This process limits the maximum stellar mass of the highest-mass star in our simulations. We call this phenomenon "fragmentation-induced starvation".

Figure 4.9 displays the accretion history of Run B. With multiple sink particles allowed to form, two subsequent sink particles build up soon after the first one, and many more follow within the next 10^5 yr. By that time the first sink has accreted $8M_\odot$. Within another $3 \cdot 10^5$ yr seven further fragments have formed, with masses ranging from $0.3M_\odot$ to $4.4M_\odot$ while the first three sink particles have reached masses between $10M_\odot$ and $20M_\odot$, all within a radius of 0.1 pc from the most massive object. Accretion by these secondary sinks terminates the mass growth of the central objects. Material that moves inwards through the disk driven by gravitational torques accretes preferentially onto stars at larger radii (Bate 2002). Eventually, hardly any gas makes it all the way to the center to fall onto the most massive objects. The diminishing of the accretion flow then allows an ionized bubble to expand, which shuts off accretion onto the most massive object entirely. It is an important point that it is not the ionizing radiation which stops the accretion, it is the subsiding accretion flow that permits the H II region to expand. This fragmentation-induced starvation prevents any star from reaching a mass greater than $25M_\odot$ in this case.

It is interesting to compare the overall accretion behavior in the multiple sink simulation Run B with a control run without any radiative feedback, Run D. In figure 4.10(a) we show the total mass in the stars M_{tot} for Run B, Run D and the single sink simulation Run A. Since the accretion is non-local in the cluster, the SFR of Run A is much lower than the SFR in Run B or Run D. But it is surprising that the SFR of Run B and Run D is virtually identical for 50 kyr. This demonstrates that the feedback by accretion heating does not influence the SFR. The turn-off around 0.68 Myr marks the point in time when the ionized bubble around the most massive star begins to expand. Since additional low-mass stars cannot form in the hot, underdense gas of the bubble, this constricts the SFR in the cluster. The evolution of M_{tot} for Run D clearly shows that there is still enough gas available to continue constant cluster growth for another 50 kyr or longer, but the gas is not allowed to collapse any more. Instead, it is swept up in a shell surrounding the H II region. Triggering could compensate this effect, but there is no star formation occuring in the swept-up shell around the bubble. Hence, the effect of the ionizing radiation on star formation in the cluster is evidently negative.

Although the accretion of all protostars from a common gas reservoir is vital to both our simulations and the competitive accretion model, there are important differences. The competitive accretion model as laid out in section 1.2.2 does not have a mechanism to prevent accretion onto the most massive stars. On the contrary, it is the major assumption of the competitive accretion model that the most massive stars will take away the gas from the lower mass stars because they reside in the center of the gravitational potential. Run B shows that the opposite behavior is found in the simulation: It is the low-mass stars that take away the gas from the massive ones and by doing so limit their growth. This is in some sense the inverse version of competitive accretion.

The relation between fragmentation-induced starvation and competitive accretion can be quantified by looking at the relation between the maximum stellar mass in the cluster, M_{max}, and the total cluster mass, M_{tot}, which is predicted by competitive accretion to follow the scaling $M_{\text{max}} \propto M_{\text{tot}}^{2/3}$. Figure 4.10(b) shows the plots for Run B and Run D. Over the whole cluster evolution, the curve for Run D lies above the curve $M_{\text{max}} \propto M_{\text{tot}}^{2/3}$, while the curve for Run B always lies below it. The general agreement of the competitive accretion prediction with the actual curves is similarly good as in Bonnell et al. (2004). This indicates that the scaling is not unique to competitive accretion, but can also be found with fragmentation-induced starvation. The figure also shows that the accretion heating suppresses low-mass star formation and that for all times the simulation with feedback contains a more massive star relative to the whole cluster mass than the simulation without feedback. When the most massive star in Run D reaches $10\,M_\odot$, much more gas is used up to create additional low-mass stars than is funneled towards the most massive one. Thus, the growth of the most massive star in Run D is much more ineffective than in Run B.

4.3.3 Disk Fragmentation

The important physical mechanism that both limits the mass growth of the individual stars in Run B and does not allow the ionizing radiation to cut off accretion in Run A is the fragmentation of the accretion disk. The fragmentation shows that it is impossible to form a $100 M_\odot$ star without companions. As soon as there is a large enough gas reservoir to form a $100 M_\odot$ star, the accretion flow of that protostar will necessarily be gravitationally unstable and fragment. This limits the gas reservoir available for each individual star. On the other hand, it also helps the ionizing radiation to escape without cutting off accretion. The dense filaments shield the ionizing radiation very efficiently, which leads to the generation of bipolar outflows. Of course, this effect is less pronounced in Run B, where much more gas is taken away from the reservoir und thus less gas is left for the filaments.

The different behavior of the filaments in the disk for the multiple sink simulation (Run B) and the single sink simulation (Run A) can be understood by looking at slices of density in the disk plane, see figure 4.11 and figure 4.12. These slices are taken at fixed coordinates, so that the movement of the filaments and the sink particles is directly visible. In particular, figure 4.12 shows how a kind of binary system forms with the sink particle being one part of the binary and a dense blob of gas being the other part. The sink particle then moves radially outwards into lower density material, but it still cannot create a bubble although it is already more massive than

4 Results

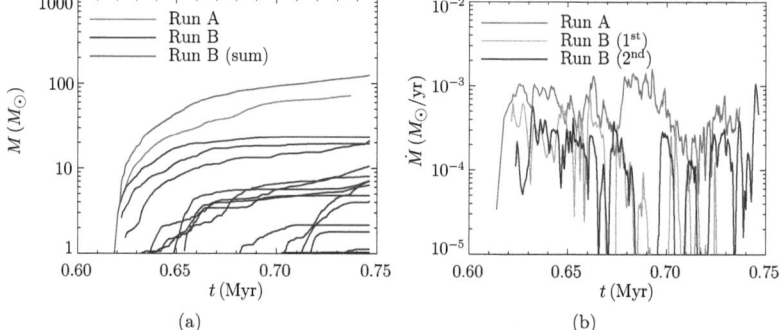

Figure 4.9: Accretion history of multi sink simulation. (a) Protostellar masses. The plot shows the growth of the individual stars in the cluster as well as the total cluster mass in Run B and the single sink Run A for comparison. Run A was stopped at $72 M_\odot$, while no sink particle in Run B exceeds $25 M_\odot$ over the simulation runtime. (b) Accretion rates. Shown are the accretion rate of the sink particle in Run A together with the accretion rates of the two most massive sink particles in Run B. The most massive stars in Run B are those which form early and keep accreting at a high accretion rate. While the accretion rate in Run A never drops below $10^{-5}\,M_\odot\,\mathrm{yr}^{-1}$, accretion can drop significantly below this value and even be stopped totally in Run B.

4.3 Collapse Simulations

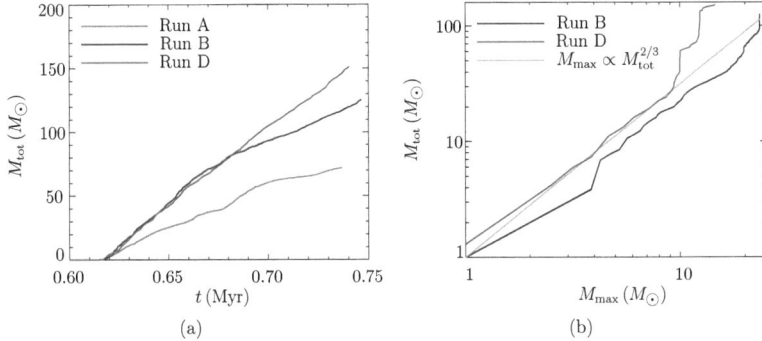

Figure 4.10: Total cluster evolution. (a) Total cluster masses as function of time. The plot shows the cluster mass M_{tot} for the multiple sink simulation Run B, the control run without any radiative feedback Run D and the single sink simulation Run A. It is interesting that the accretion heating does not change the SFR at all. The turn-off signalizes the expansion of the ionized bubble, which prevents the formation of further low-mass stars and does not trigger any star formation. (b) M_{tot} as function of M_{max}. We plot the curves for Run B, Run D as well as a curve with $M_{\mathrm{max}} \propto M_{\mathrm{tot}}^{2/3}$ as predicted by competitive accretion (see section 1.2.2). The simulation with radiative feedback always lies below this curve, whereas the contol run always lies above it.

$70 M_\odot$. One can also see that a filament passes directly through the sink particle, which illustrates the strength of the accretion flow.

In Run B, the filaments in the disk form secondary sink particles, which remove gas from the common reservoir. Consequently, the most massive stars cannot attain their high accretion rates, and it is this halting of the accretion flow which allows the ionizing radiation to create a bubble around the massive protostars. Figure 4.11 demonstrates that it takes a few 1000 yr for the bubble to develop. First, it can only blow a semi-circle into the disk since the accretion flow is still strong enough in the opposing direction to prevent a symmetric expansion. Only when the accretion flow continues to decline, the ionized bubble can isolate the protostar from the disk. At this point, accretion onto the most massive star stops entirely. The second most massive protostar never is strong enough to create an isolated bubble, but as the first bubble expands, the second most massive star enters the bubble and then helps to power the expansion.

In contrast, the sink particle in Run A is embedded in an accretion flow at all times. The ionizing radiation does blow away a significant fraction of gas from the sink particle, but it is not able to do so simultaneously in all directions. There is always a region around the sink particle where the gas is dense enough such that evaporation by the ionizing radiation is not sufficiently strong to stop the accretion flow.

4.3.4 Bipolar Outflows

The differences between Run A and Run B are also evident from the characteristics of the bipolar outflows. Figures 4.13 and 4.14 show density slices perpendicular to the disk plane for both simulations. The slices are comoving with the (most massive) sink particle, so that the protostar is always in the plane of the slice, and the star is kept at the center of the image. The scales and times of the images are the same as in figures 4.11 and 4.12, respectively. The outflows are driven by the same general mechanism. As the filament approaches the sink particle, the ionizing radiation is shielded. Generally, the filament approaches the protostar either slightly above or below the midplane, so that the radiation is shielded above or below the disk plane. The hot gas there can then recombine and cool down at a short timescale. The thermal support gets lost, and the shell of the outflow falls back towards the disk because of its gravitational attraction. Since the accretion flow is chaotic, these ionization and recombination events can alternate on times as short as 100 yr. A really large outflow requires either constant support by the ionized gas pressure or fluctuations rapid enough that the shell has no time to fall back. This explains why the larger outflows are mostly one-sided.

Another feature common to both simulations is the dynamics inside the shell and, in particular, the instability of the shell itself. As the ionizing radiation evaporates parts of the disk and the approaching filaments, strong shocks and rarefaction waves run through the cavity of the outflow. These waves are initially subsonic in the ionized gas, but when the ionized gas recombines, they continue to propagate supersonically through the neutral outflow. When these waves hit the shell, they drive instabilities

4.3 Collapse Simulations

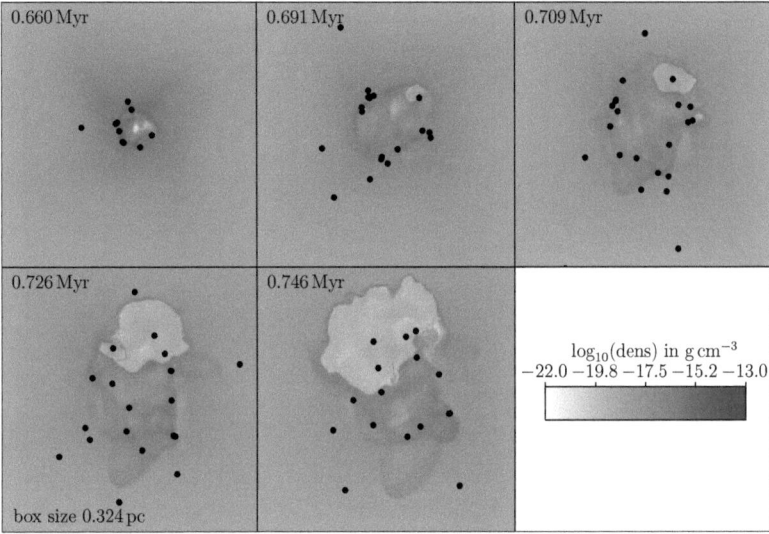

Figure 4.11: Disk fragmentation in the multiple sink simulation (Run B). The figure shows the gas density in slices in the midplane of the disk. The filaments in the disk form a successively growing number of protostars. As the gas reservoir around the massive protostars is exhausted, the thermal pressure of the ionized gas creates a bubble around the star. This stops the accretion process. In the course of the simulation, the bubbles grow in size and finally merge.

4 Results

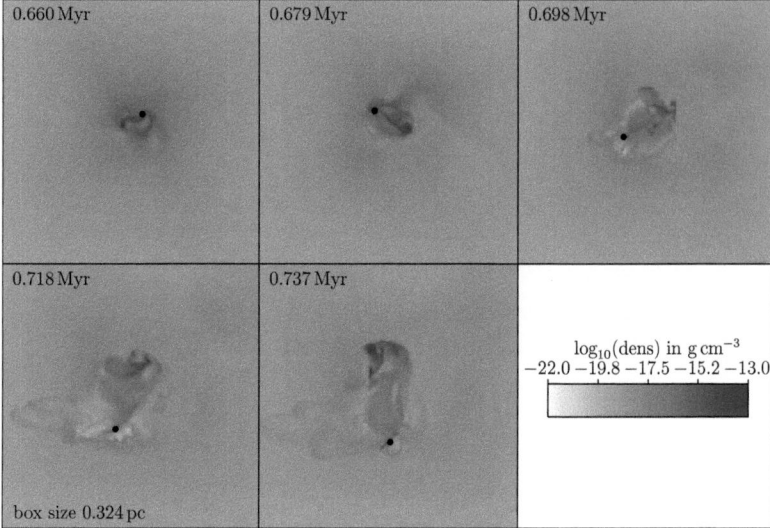

Figure 4.12: Disk fragmentation in the single sink simulation (Run A). The images show the same region as figure 4.11. This time the filaments do not form protostars but blobs of gas which are puffed up by the Jeans heating. The ionizing radiation is unable to create a bubble in the strong accretion flow large enough to stop accretion.

that can grow to finger-like structures. These fingers are very similar to the instabilities discussed in section 2.4.6, but the formation of the fingers in the shell is driven by the dynamics inside the shell, so it is not a pure ionization front instability. Nevertheless, the instability is certainly amplified because of the fact that the shell expands into a density gradient, similar to the conventional instabilities.

Figure 4.13 shows the outflows driven by the most massive sink particle in Run B. The efficient shielding by the filaments and the motion of the sink particle hinder the thermal pressure of the ionized gas to drive a symmetric bipolar outflow at early times. The shielding of the ionizing radiation leads to a drop in the thermal pressure which drives and supports the outflow, and thus the outflow can be quenched again. At later times, the strong accretion flow onto the sink particle stops, and this allows the sink particle to drive a much larger outflow, which in the end has the form of a bubble. Initially, this bubble does not expand spherically from the protostar. Since the gas in the disk is denser, it constitutes a barrier for the shell that it cannot push away so easily. But the bubble is also not symmetric with respect to the disk plane. The lower lobe grows much faster than the upper lobe. This is because the ionizing radiation already had blown an outflow on the lower side at the point when the accretion flow stopped, whereas the upper side was shielded by the accretion flow. When the accretion flow stopped, the bubble first had to overcome this delay. In the end of the simulation, the bubble is much more spherically shaped. The ionizing radiation preferentially evaporates the gas closest to the source, where the heating is the strongest.

The situation for the sink particle in Run A is presented in figure 4.14. The general properties of the outflow are very similar, but the big difference is that here the accretion flow never stops. On the contrary, figure 4.9 shows that the accretion rate increases towards the end of the simulation. This strong accretion flow fully absorbs the ionizing radiation, and the sink particle is unable to build up a bubble although it has more than three times the mass of the most massive star in Run B. This is even more surprising since the sink particle moves far away from the center of the rotationally flattened structure into less dense material. The figure demonstrates that the accretion disk has faded in the last frame.

We can compare the model with observations of young massive stars by computing observable line and continuum emission using the radiative transfer code MOLLIE[1]. Figure 4.15 shows such a comparison at a time when a $20 M_\odot$ star has formed in Run B. The star lies at the center of a dense accretion disk that is one of several within the larger-scale rotationally flattened flow. In the simulated observations, the accretion disk is identifiable by the bright $NH_3(3,3)$ emission and by the signature of rotation, red and blue-shifted line velocities on either side of the star. The larger-scale, rotationally-flattened flow that feeds the accretion disk is indicated by the nearly horizontal contours of the line emission and also the weaker velocity gradient across the whole image.

At this stage in the hydrodynamic simulation the star is hot enough to begin to

---

[1]The analysis presented here was carried out in collaboration with Eric Keto.

4 Results

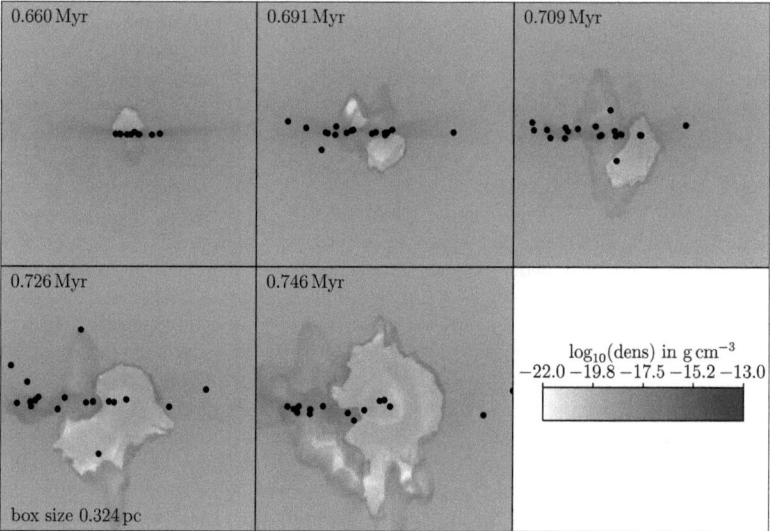

Figure 4.13: Bipolar outflows in the multiple sink simulation (Run B). The slices show the gas density in a plane through the most massive sink particle perpendicular to the disk. In the beginning of the simulation, the outflow is bipolar but not symmetric because the filaments prevent the thermal pressure to drive simultaneously both lobes of the outflow. This effect fades when the accretion flow drops, and the sink particle can even blow an expanding bubble although it does not grow in mass anymore in the last three frames shown.

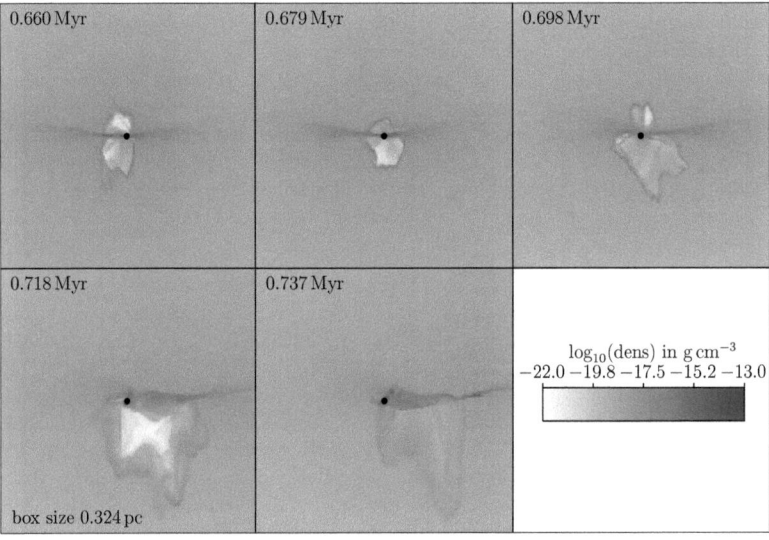

Figure 4.14: Bipolar outflows in the single sink simulation (Run A). The images show that the size of the outflow is not directly related to mass of the protostar. Instead, it is given by the time the thermal pressure has had to steadily drive it. If the thermal support is lost due to recombination, the outflow gets quenched again. This is apparent in the last frame, where the star is the most massive but there is no visible cavity around it.

4 Results

ionize its own accretion flow. The ionized gas is expanding off the accretion disk and in the direction down the steepest density gradient of the molecular gas, perpendicular to the disk and rotationally-flattened flow. The ionized gas is visible in the simulated observations of the 1.3 cm continuum as an H II region just above the accretion disk. Although the H II region appears spherical, we know from the numerical hydrodynamical simulation, and also from the simpler analytic model (Keto 2007), that the H II region is a conical-shaped outflow driven by thermal pressure down the density gradient maintained by the gravitational field of the star at the base of the outflow. The ionized outflow is continuously supplied by photoevaporation of the accretion disk, which is itself re-supplied by the larger-scale molecular flow. Because the ionized outflow derives from the rotating accretion disk, the gas flow in the H II region has components of both rotation and outflow resulting in an outward-twisting spiral flow. An observation that only partially resolves the spatial structure of the ionized flow sees a velocity gradient oriented in the direction between that of the disk and the outflow. This is shown in a simulated observation of the H II region in the H53α radio recombination line and 1.3 cm continuum (see figure 4.15, upper right panel). The orientation of the apparent velocity gradient depends on the relative speeds of the rotation and outflow.

Observations of the W51e2 region, thought to contain a massive protostar of about $20 M_\odot$, show these several features (figure 4.15, lower panels). These observations are more fully discussed by Zhang et al. (1998) and Keto & Klaassen (2008). In the lower left panel of figure 4.15, an accretion disk is identifiable in $NH_3(3,3)$ line brightness and velocities. The disk orientation is from the south-east (red velocities) to the north-west (blue velocities) at a projection angle of 135° east of north (counter-clockwise). Just off the mid-plane of the accretion disk, an H II region is seen in the 1.3 cm radio continuum. The $NH_3(3,3)$ in front of the H II region is seen in absorption and redshifted by its inward flow toward the protostar. The lower right panel of figure 4.15 shows the velocity of the H53α recombination line (Keto & Klaassen 2008) in the H II region. The direction of the velocity gradient between the directions of rotation and the outflow indicates that the ionized gas has both velocity components and spirals outward off the disk

The hydrodynamic simulation is not intended to model the W51e2 region. For example, the spatial scales are not identical. Nonetheless, the comparison illustrates a few points about the star formation in W51e2. The velocity gradient and red-shifted absorption observed in the NH_3 lines indicate that the massive star is forming within a molecular accretion disk and large-scale inflow. The offset of the H II region from the dense molecular gas in the accretion flow indicates that the ionized gas is expanding asymmetrically and perpendicular to the rotational flattening of the molecular flow. The misalignment of the velocity gradients in the molecular and ionized flows indicates that the H II region is a spiral outflow of ionized gas. The apparent size of the H II region is a consequence of the density gradient in the ionized flow. Therefore the age of the H II region is commensurate with the time scale of the accretion flow rather than the much shorter sound-crossing time of the H II region.

4.3 Collapse Simulations

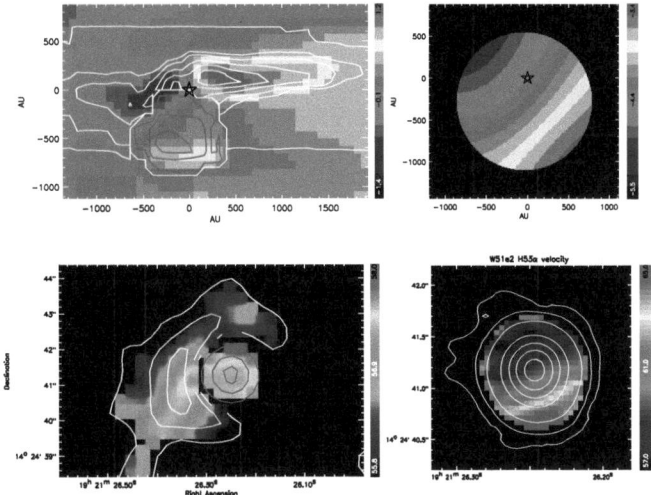

Figure 4.15: Ionized accretion flow in observation and simulation. Comparison of line and continuum emission simulated from the model (upper panels) and actually observed from the W51e2 region (lower panels). The left panels show the $NH_3(3,3)$ line emission strength in white contours, the molecular line velocities as the background color, and the 1.3 cm free-free continuum from ionized gas in red contours. The right panels show the H53α recombination line velocities from the ionized gas. The simulated H53α observations (upper-right) are convolved to a spatial resolution of 750 AU (FWHM) to better match the resolution of the actual observations (lower-right). In the simulated observations, the origin of the coordinates is the $20 M_\odot$ protostar at the center of the accretion flow as marked. The accretion flow and disk are viewed edge-on. The spatial scale in the observations (lower panels) is 7000 to 8000 AU per arc second assuming a distance of 7 to 8 kpc to W51e2. The color bar on the right of each figure shows the molecular velocities in $km\,s^{-1}$, including the LSR velocity of W51e2, approximately $57\,km\,s^{-1}$. The white contour levels are 50 % through 90 % of the peak brightness temperature of 71 K (upper left) and 0.1, 0.2 and 0.3 Jy beam^{-1} km s^{-1} (lower left). The red contour levels are 30 %, 70 %, and 95 % of the peak 1.3 cm continuum brightness temperature of 8902 K (upper left) and 0.07, 0.14, and 0.22 Jy beam^{-1} (lower left). The white contours on the H53α observations (lower-right) show the 7 mm continuum emission at 2, 4, 10, 30, 50, 70, and 90 % of the peak emission of 0.15 Jy beam^{-1}. The molecular line observations are from Zhang et al. (1998), the H53α observations from Keto & Klaassen (2008), both having a beam size of about $1''$. Coordinates are in the B1950 epoch.

4.3.5 H II Region Morphologies

The dynamics of the H II region is also reflected in the radio continuum maps generated from the simulation data. Unless stated otherwise, the radio maps are generated for VLA parameters at a wavelength of $\lambda = 2$ cm with a full width at half maximum of the beam $0''\!.14$ and a noise level of 10^{-3} Jy (see table 2.1). The assumed distance is 2.65 kpc.

The continuum maps show the continuous build-up and destruction of UC H II regions. The timescale for large-scale (~ 5000 AU) changes can be as short as ~ 100 yr. This flickering is caused by the accretion flow in which the sources are embedded. When filaments in the disk are accreted by the protostar, the ionizing radiation is effectively shielded, so that the gas above the filament can recombine and cool down. Since the accretion flow is chaotic, the interplay between the radiation feedback and the infalling material results in highly stochastic ionization and recombinaten processes in the surrounding gas.

This effect is demonstrated in figure 4.16. It shows some dramatic changes in the H II region around the most massive star in Run B. Between $t = 0.6592$ Myr and $t = 0.6595$ Myr (within 300 yr), an area with a diameter of ~ 6000 AU suddenly recombines. Changes like this not only affect the physical size of the H II region, but they can also alter their morphology. From $t = 0.6668$ Myr to $t = 0.6671$ Myr (again within 300 yr), the morphology of the UC H II region surrounding the most massive protostar changes from shell-like to core-halo because of a large-scale recombination event that clears the rim of the shell. The shielding by the filaments also controls how ionizing radiation can escape perpendicular to the disk. This reverses the cometary H II region around the star between $t = 0.6524$ Myr and $t = 0.6534$ Myr (within 1000 yr). The three examples given in figure 4.16 indicate that the morphology of H II regions around accreting massive protostars depends sensitively on accretion events close to the protostar.

The flickering observed in the simulations also resolves the long-standing lifetime problem for UC H II regions (see section 1.1.4). Since UC H II regions are not freely expanding bubbles of gas which monotonically increase in size, their diameter cannot be related to their age. Because H II regions embedded in accretion flows are continuously fed, and since they flicker with variations in the flow rate, their size does not depend on their age until late in their lifetimes. An extreme version of the discrepancy between protostellar mass and size of the H II region occurs in Run A, where the $70 M_\odot$ protostar has almost no visible H II region. It is totally quenched by the strong accretion flow.

While the source in Run A never stops accreting, the most massive stars in Run B finally stop growing when the gas reservoir around it is fully exploited. The ionizing radiation then creates a bubble around it which later also encloses the second most massive star. As soon as the bubble is created, the flickering around the most massive star stops, and the emission becomes much fainter than before and extends over a great fraction of the cluster size. Their H II regions merge into a compact H II region, the type that generally accompanies observed ultracompact H II regions (Kim & Koo

4.3 Collapse Simulations

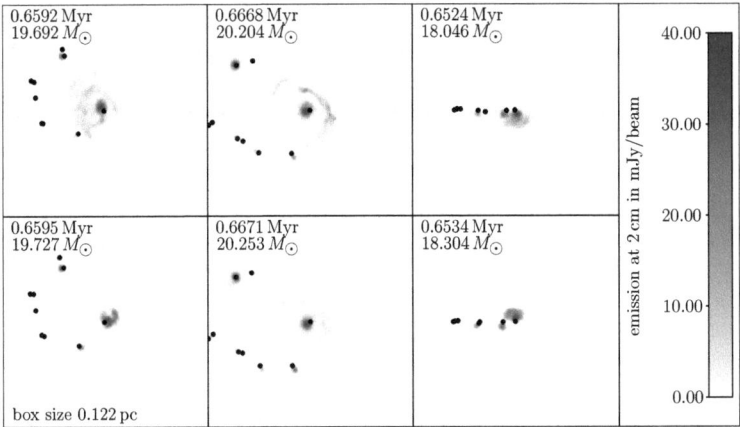

Figure 4.16: Changes of H II regions. The frames show images from H II regions around the most massive protostar in Run B. The frames in the lower panels show the H II region at a slightly later time than the images in the upper panels. The left-hand panels show the recombination of ionized gas with a diameter of $\sim$ 6000 AU within 300 yr. The midway panels demonstrate how the morphology of the H II region changes from shell-like to core-halo in 300 yr. The right-hand panels present the reversal of a cometary H II region in 1000 yr. The box size displayed is 0.122 pc. The mass of the protostar is shown in each frame. The images are synthetic 2-cm VLA observations at 2.65 kpc.

2001). We stress again that it is not the ionizing radiation which stops the accretion flow by creating a bubble, rather it is the decaying accretion flow which allows the bubble to expand.

The extended H II regions found in the simulations show a large amount of substructure. Equation (2.145) shows that the emission from free-free transitions scales with the square of the number density of free electrons, n_e. This explains the emission peak close to the protostar, where very dense gas in the accretion flow gets partially ionized. However, not all emission peaks are associated with stars.

Figure 4.17 shows some examples. The upper left panel shows an H II region with a shell-like structure. The shell clearly exhibits a peak on its rim that is several 1000 AU away from any nearby star. The shell is created by dense shocks running through the H II region, which are replenished by material from the accretion flow. The emission of this dense gas is what creates the shell. Another example is shown in the lower left panel of figure 4.17. It shows a dense blob of gas which is irradiated by a massive star and creates a peak that looks like as if it was indicating the position of a second star. Obviously, peaks in emission maps are not an ideal hint to the potential coordinates

4 Results

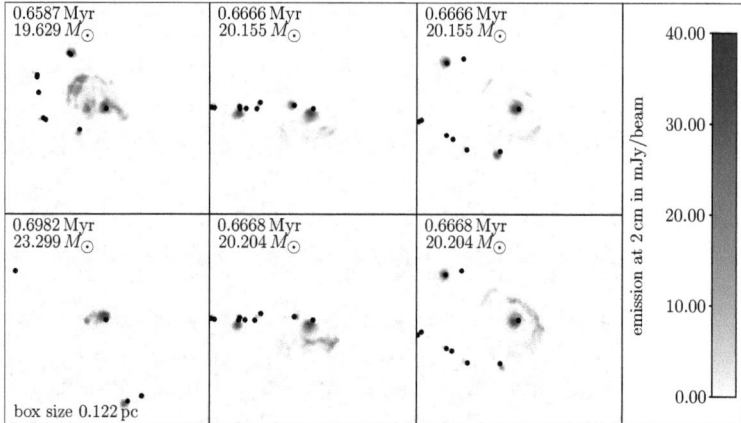

Figure 4.17: Emitting structures of H II regions. The frames show images from H II regions around the most massive protostar in Run B. The left-hand panels give two examples of emission peaks that are not directly related to stars. This questions the usual understanding that coordinates of stars should be directly related to emission peaks. The middle panels show an edge-on view of the star cluster with a cometary H II region around the most massive star. Within 200 yr, a shock is launched at the protostar, runs through the H II region and creates a filamentary structure. The right-hand panels show the situation face-on. Here, the shock looks like a shell expanding from the protostar. This indicates that the appearance of H II regions is closely related to accretion events close to the protostar. The images are synthetic 2-cm VLA observations at 2.65 kpc.

of stars.

The aforementioned shocks contribute largely to the emission seen in the emission maps. The middle panels in figure 4.17 show an edge-on view of the rotationally flattened structure of the star cluster. The upper panel shows that the most massive star has created a cometary H II region. The lower panel shows the same region 200 yr later. The ionizing radiation has blown away gas from the accretion flow close to the protostar. This shock runs away from the star and creates a filament of strong emission across the H II region. The right-hand plots in figure 4.17 show the same region face-on. From this viewing angle, the shock shows up as shell-like structure. This shows that the shell does not trace the edge where the ionizing radiation hits the accretion disk, but rather shocks generated from inflowing gas. The shell-like structures around accreting protostars can be interpreted as indirect evidence for the accretion process.

The origin of the shell morphology changes when accretion ceases. Figure 4.18 shows the late-stage evolution of the star cluster. The most massive star has stopped

4.3 Collapse Simulations

accreting and begins to blow a bubble into the ambient gas. The left-hand plots show this bubble face-on (upper panel) and edge-on (lower panel). Here, the strong shell-like emission cleary comes from the dense gas in the rotationally flattened structure around the protostar and not from a shock launched by the protostar.

While the accretion onto the most massive star, which created the bubble, has stopped and cannot directly affect the structure of the H II region anymore, it can still be influenced by other stars which interact with the gas. Two such events occur in Run B and are shown in figure 4.18. The middle panels show a time sequence of a $7.8 M_\odot$ star approaching the rim of the shell. Its ionizing radiation is strong enough to create sufficient thermal pressure to blow away the rim of the shell from its direct neighborhood. The right-hand panels show the same star 8200 yr later, when it has already entered the bubble. Now its ionizing radiation can freely expand into the bubble. The gravitational attraction of the star is strong enough to pull along a dense stream of gas. This stream allows the star to grow in mass although it has entered the bubble filled with underdense gas. This suggests that deformed shells could be indicative of stars inside large-scale H II regions.

The simulation data presented here offer the unique opportunity to look at the same H II region from different viewing angles. We have already seen that the observed morphology depends crucially on the position of the observer. We investigate this observation in detail for some H II regions appearing in Run B.

Figure 4.19 displays an H II region around the most massive star in Run B. The first panel shows the region face-on, and the successive panels rotate the region around an axis in the plane of the rotationally flattened structure by 18° until a quarter rotation of 90° is reached. Hence, the last panel shows the region edge-on. This sequence of images demonstrates how a shell-like morphology can change into a cometary morphology. The transition angle at which the shell-like morphology turns into a cometary morphology is about 72° in this case.

However, we will get a different result if we perform the rotation around a different axis. Figure 4.20 starts with the last frame of figure 4.19 and successively rotates the region by 90° around the polar axis. This means that the view is edge-on for all times. At an angle around 36°, the cometary region develops strong shell-like structures. We already know that these filaments have their origin in gas blown away by the protostar. This region is of shell-like type since it is bounded by a dense filament. This means that shell-like regions can occur both at face-on and at edge-on view. In principle, the same holds true for cometary regions. They can also be observed face-on when the ionizing radiation is shielded anisotropically.

We have repeated the same exercise for different H II regions and find that the transition angle between shell-like and cometary morphologies varies a lot. Figure 4.21 shows two more rotations that transform a face-on view into an edge-on view. We show only the first and last images as well as the transition angle. The upper panels show the transformation of a shell-like morphology into a cometary morphology at $t = 0.6864$ Myr, the lower panels show a similar transition at $t = 0.6925$ Myr. The central star has a mass of $22.532 M_\odot$ and $23.025 M_\odot$, respectively. The transition angle is about 36° in the former and 54° in the latter case. They depend on the details of

4 Results

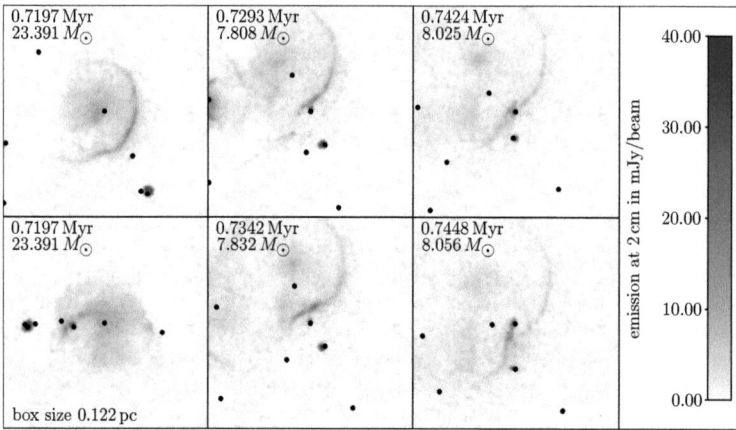

Figure 4.18: Bubble created by the most massive protostar in Run B. The left-hand panels show this bubble face-on (upper panel) and edge-on (lower panel). The emission in the shell has its origin in the rotationally flattened structure around the protostar and not in a shock running through the H II region. The middle and right-hand panels show a time sequence of a second star interacting with the dense gas that bounds the bubble. In the middle panels, this star blows away a dense filament by its own ionizing radiation. The right-hand panels demonstrate what happens when it enters the bubble. By its gravitational field, it pulls a dense stream of gas into the bubble behind itself. Broken-up shells could thus be a helpful observational signature to locate stars inside H II regions. The images are synthetic 2-cm VLA observations at 2.65 kpc.

4.3 Collapse Simulations

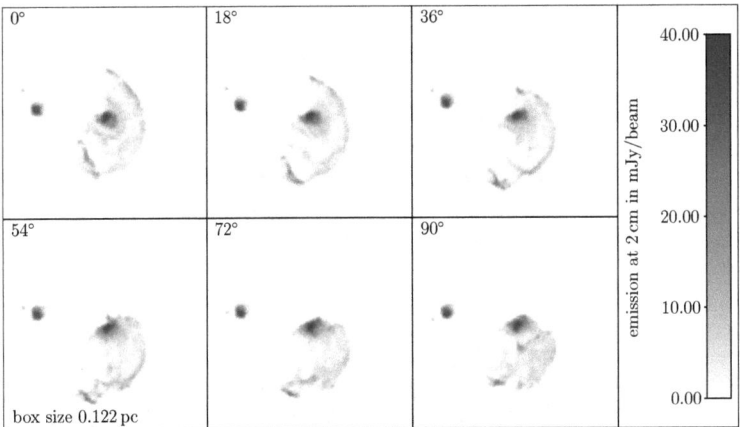

Figure 4.19: Rotation from face-on to edge-on view of an H II region in Run B around a star with $22.956 M_\odot$ at $t = 0.6907\,\text{Myr}$. The view on the first panel is face-on. The further panels show a successive rotation by 18° around an axis in the plane of the rotationally flattened strucutre. The last panel shows the region edge-on. At an angle of about 72°, the morphology has turned from shell-like into cometary. The images are synthetic 2-cm VLA observations at 2.65 kpc.

the structure. A universal angle above which transition occurs does not exist.

Morphologies of UC H II regions as they are found in sites of massive star formation can be classified by their type. Wood & Churchwell (1989) and Kurtz et al. (1994) used the types shell, cometary, core-halo, spherical, irregular and unresolved. De Pree et al. (2005) abandoned the core-halo morphology and introduced a new bipolar category for elongated H II regions. The reason given for abandoning the core-halo morphology is that most H II regions are surrounded by faint emission, which produces a halo around any H II region. Though this may be true, we find it useful to keep this morphological type for regions with a pronounced central peak and a fainter envelope. The presence of a pronounced envelope clearly distinguishes this morphology from the spherical type, which we also find.

In addition, we do not require shell-like regions to be not centrally peaked. Although the larger, late-time shells do indeed not have central peaks, the UC H II regions associated with accreting protostars do have central peaks because they ionize their own accretion flow. In fact, observations with a very high sensitivity and resolution do find centrally peaked shells that were previously classified as spherical (Rodríguez et al., 2003). We predict that more regions of this type will be found as soon as observations with better resolution and sensitivity become available also for massive star forming regions that are farther away. Figure 4.22 demonstrates the importance of resolution

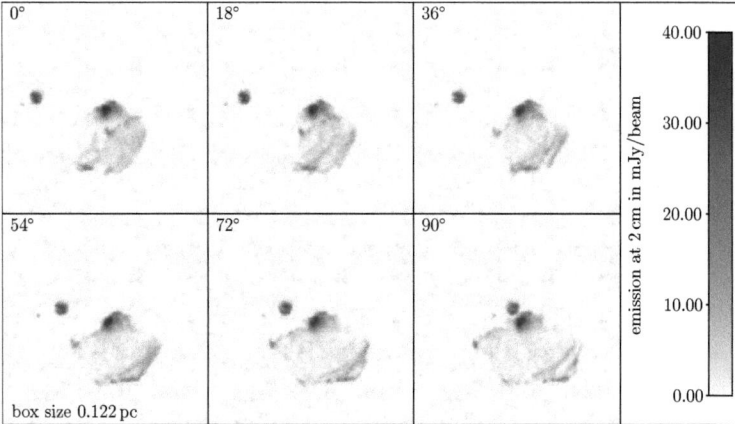

Figure 4.20: Rotation around polar axis of an H II region in Run B around a star with $22.956 M_\odot$ at $t = 0.6907$ Myr. The first panel is identical with the last panel of figure 4.19. The region is successively rotated around the polar axis, so that the view is edge-on for all angles. The morphology changes from cometary to shell-like at about 36°. The images are synthetic 2-cm VLA observations at 2.65 kpc.

4.3 Collapse Simulations

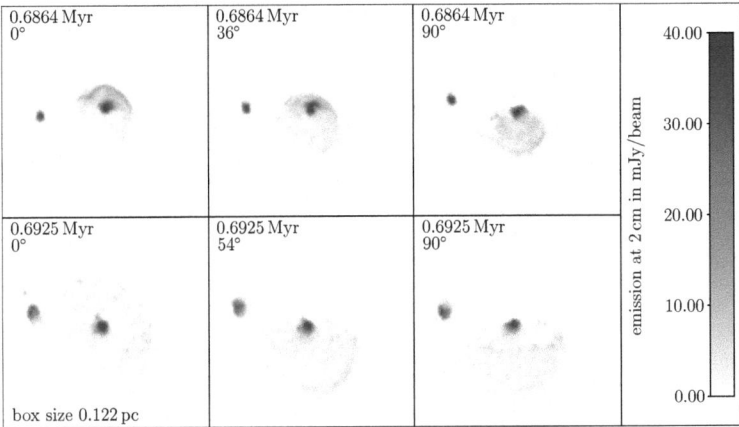

Figure 4.21: Different transition angles from shell-like to cometary H II regions in Run B. The upper panels show an H II region at $t = 0.6864$ Myr around a $22.532 M_\odot$ star, the lower panels at $t = 0.6925$ Myr when the star has $23.025 M_\odot$. The polar transition angles are $36°$ and $54°$, respectively. The images are synthetic 2-cm VLA observations at 2.65 kpc.

and sensitivity in identifying the correct morphology. The figure shows synthetic maps of the same shell-like H II region for VLA parameters at 2 cm for different distances to the observer. The shell disappears between 6 and 10 kpc, where both the spatial resolution and the noise level make it impossible to distinguish between parts of the H II region and pure noise. On the other hand, the amount of centrally peaked shells in the simulation may be reduced by including additional feedback processes like line-driven stellar winds or magnetically driven outflows, which might be able to create a cavity around the protostar. If these processes would be strong enough to thin out the dense accretion flow sufficiently to remove the peaks remains an open question.

The new bipolar type is not well defined. De Pree et al. (2005) gave only one example for this new category where the bipolar shape is not very distinctive. Churchwell (2002) required an hourglass shape for the bipolar morphology which is not present in their example. Although we do find morphologies that look bipolar in the simulations, there are only very few of them, and their features are not very pronounced. The small number of bipolar regions is in agreement with observations (Churchwell 2002, De Pree et al. 2005). All of the bipolar regions could equally well fall into one of the other classes, which is why we do not take this category into account.

The sensitive dependence on viewing angle and the high time variation of the H II region caused by the accretion flow is sufficient to produce morphologies of any mentioned class in a single simulation. Figure 4.23 shows maps from Run A with only one ionizing source. The displayed shell-like and core-halo morphologies are face-on views,

4 Results

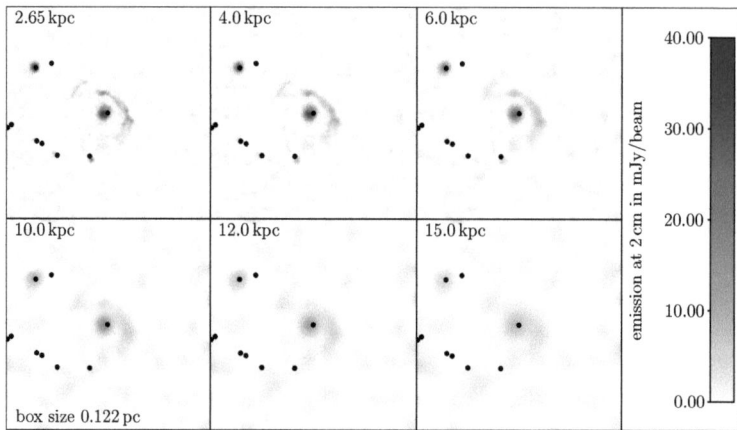

Figure 4.22: Shell-like H II region of Run B with a central peak for different distances to the observer. The synthetic 2-cm VLA maps show that the shell-like feature disappears between 6 and 10 kpc. To detect it at such high distances, observations with higher spatial resolution and sensitivity are required.

whereas the cometary H II region is viewed edge-on. As discussed above, the same morphologies can equally well be obtained at different viewing angles. It is apparent that the size of the H II region does not scale with the mass of the protostar. On the contrary, the irregular region corresponds to the largest protostar, but it is among the smallest H II regions. Figure 4.24 shows some examples for the different morphologies in Run B.

For comparison with observational surveys (Wood & Churchwell 1989, Kurtz et al. 1994), we generate a census of morphologies in Run A and Run B. We select 25 snapshots from each simulation and 20 randomly chosen viewing angles. This gives a total amount of 500 images per simulation. The different viewing angles take into account the fact that due to the special geometry of the setup, some morphologies are preferentially found at different orientations. For example, shell-like morphologies are found mostly face-on and cometary morphologies edge-on. The set of different viewing angles, uniformly distributed on the unit sphere around the center of the computational domain, avoids statistical biases by this effect. By the same token, the distribution of morphologies also changes with time. Since many stars in Run B reach a mass of $10\,M_\odot$ by the end of the simulation, the later times contain more spherical and unresolved H II regions than the beginning of the simulation. Since we do not know the geometry and the evolutionary stage of the UC H II regions in the surveys, we assume randomly distributed orientations and ages and thus average over different viewing angles and simulation snapshots to get a representative sample.

4.3 Collapse Simulations

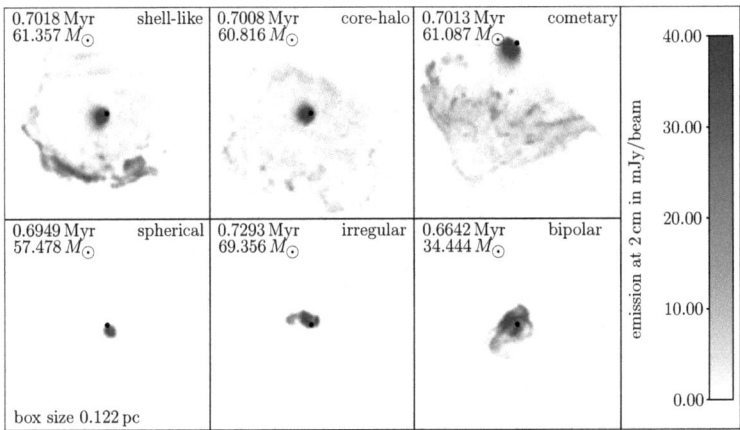

Figure 4.23: Different morphologies observed in Run A with a single ionizing source only. All morphologies found in surveys are present, depending on the time in the simulation and the viewing angle. The lower right panel demonstrates the problem with the bipolar category. The region is clearly elongated, but it also shows a shell-like structure. Since elongation alone seems to be insufficient to define a category on its own, we do not consider bipolar regions as a separate category. The images are synthetic 2-cm VLA observations at 2.65 kpc.

4 Results

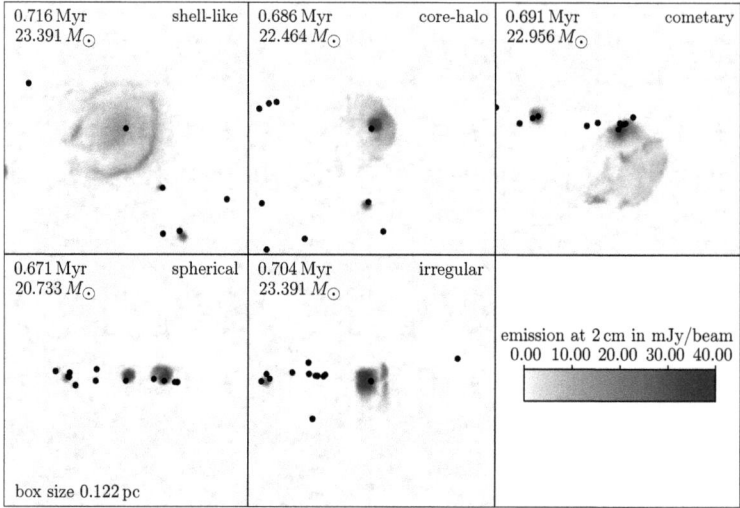

Figure 4.24: Different morphologies observed in Run B. This figure shows UC H II regions around massive protostars in Run B at different time steps and from different viewpoints. The protostellar mass of the central star which powers the H II region is given in the images. The H II region morphology is highly variable in time and shape, taking the form of any observed type (Wood & Churchwell 1989, Kurtz et al. 1994) during the cluster evolution. The images are synthetic 2-cm VLA observations at 2.65 kpc.

Type	WC89	K94	Run A	Run B
Spherical/Unresolved	43 %	55 %	19.3 %	59.4 %
Cometary	20 %	16 %	6.5 %	3.8 %
Core-halo	16 %	9 %	14.6 %	3.9 %
Shell-like	4 %	1 %	3.1 %	6.4 %
Irregular	17 %	19 %	56.5 %	26.6 %

Table 4.2: UC H II region morphologies in surveys and simulations. The table shows the morphology statistics of UC H II regions in the surveys of Wood & Churchwell (1989) (WC89) and Kurtz et al. (1994) (K94) as well as from a random evolutionary sample fron Run A and Run B of 500 images for each simulation. The statistics for Run A is in disagreement with the observations.

To achieve consistency with the observations, we use contour plots to identify the morphological classes and follow the definitions given in Wood & Churchwell (1989) (see also table 1.2).

The results of this statistical analysis are presented in table 4.2. Given the deviations amongst the observational surveys, the statistics from the multiple sink simulation Run B is very close to the observational numbers. In particular, we find that roughly half of the sample represents spherical or unresolved UC H II regions, in agreement with the classical lifetime problem. However, the statistics also shows that the relative numbers of spherical and unresolved H II regions in Run A disagree by more than 20 % with the observational findings. This clearly rules out theoretical models in which massive stars form alone. Since the protostar in Run A grows very quickly, it cannot generate such a large number of strongly confined H II regions. Instead, lots of irregular H II regions form. The only way to get a high number of spherical and unresolved H II regions is the formation of a stellar cluster. This again shows that Run B is a much more realistic model for massive star formation. The results from table 4.2 demonstrate that morphology statistics are a useful diagnostic tool to distinguish different models of massive star formation.

Another interesting property of UC H II regions is their SED. As explained in section 2.5.1, the SED typically grows as ν^2 in the optically thick regime (small frequencies) and falls off like $\nu^{-0.1}$ in the optically thin regime (large frequencies). This behavior is illustrated in figure 2.4. However, many observed SEDs of H II regions do not behave in this simple manner but show abnormal scaling exponents (Lizano 2008). In particular, there are SEDs which grow with an intermediate exponent of ν^1 over a frequency interval that can be as large as the whole VLA band coverage (see table 2.1). Abnormal scaling exponents can be reproduced by H II region models with ionized gradients (Panagia & Felli 1975, Olnon 1975, Franco et al. 2000, Avalos et al. 2006, Keto et al. 2008), hierarchical clumps (Ignace & Churchwell 2004) and additional dust emission (Rudolph et al. 1990, Pratap et al. 1992, Beuther et al. 2004, Keto et al. 2008).

4 Results

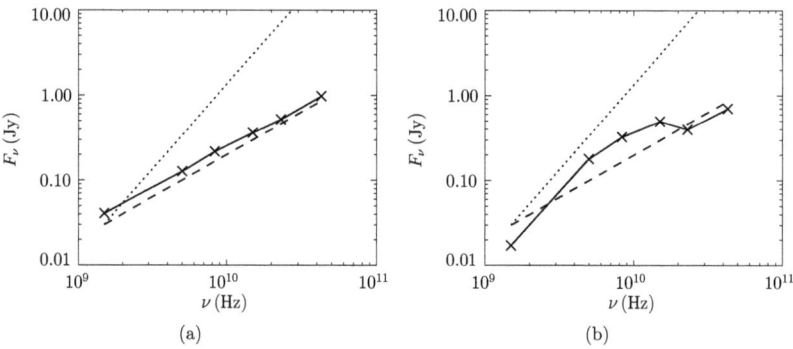

Figure 4.25: Abnormal SEDs in Run B. The images are generated for the VLA bands L to Q assuming a distance of 2.65 kpc. The dotted line grows with ν^2, the dashed line with ν^1. (a) SED that grows with ν^1 across all six bands. This slope can be a combined effect of ionized gadients and shadowing by clumps. (b) SED that has a pronounced minimum at 23 GHz. The rising at high frequencies is produced by a very bright core of the UC H II region. There is no clear relation between SEDs and morphologies.

The data from the numerical simulations is certainly more realistic than the simple ionized gradients or hierarchical clumps models, but the postprocessing only takes free-free radiation into account, not dust emission. But even without dust, we can get very complicated SEDs as well as SEDs that have a uniform scaling exponent around 1. We show two examples in figure 4.25. They are generated from synthetic VLA observations from bands L to Q of UC H II regions from Run B assuming a distance of 2.65 kpc. Figure 4.25(a) shows an H II region with a constant growth of ν^1 across all observed bands. This is exactly the kind of abnormal SED found in observations. The result can be a combined effect of density gradients in the ionized gas as well as shadowing by clumps, just as in the simpler analytical models. Figure 4.25(b) displays an SED that first grows with ν^2 and then turns over as expected, but then has a minimum at 23 GHz and starts growing again. It is interesting that we do not need any dust to find such an emission excess at high frequencies. The images reveal that the envelope fades away at high frequencies, but the core gets much brighter and produces the overshoot. We have found several other SEDs with this curious shape, and most of them have the minimum at the same frequency band and show a bright core for high frequencies. Both kinds of SEDs can be found in many different H II regions with various morphologies. There is no clear relation between morphologies and the form of the SED.

4.3.6 Time Variability of UC H II Regions

Since the H II regions flicker on a time scale of only 10 yr, a comparison with observed changes in H II regions is possible[2]. In particular, we can try to learn about the curious shrinking H II regions mentioned in section 1.1.4.

For this time series comparison at high temporal resolution, we select a few time intervals in the multiple sink simulation (Run B) in which the continuum emission from a particular H II region is variable, and with that H II region well isolated in space to be able to measure its individual properties. We then produce synthetic 2-cm continuum maps with VLA parameters at every $\sim 10\,\mathrm{yr}$ from the simulation output. From this data, we can determine the observational scale length H for the H II region, which is defined as the equivalent diameter of a circle with the same area as the emission. We also study the evolution of the total flux $F_{2\,\mathrm{cm}}$ over the H II region and the accretion rate $\dot{M}$ of the sink particle that powers the H II region.

Figure 4.26 shows two such events at high temporal resolution. The interpretation of figure 4.26(a) is simple. The flux and size of the H II region go down exactly at the same time when the accretion rate of the sink particle goes up. Obviously, a dense blob of gas is being accreted, which shields the ionizing radiation and therefore causes the H II region to shrink. Once this dense gas is gone, the star begins to ionize its environment again, so the H II region starts growing. But this expansion needs much more time than the shrinking since it is governed by gas-dynamical processes, whereas the contraction is caused by recombination.

Figure 4.26(b) is much more tricky to understand because there seems to be a delay between the accretion process and the shrinking of the H II region. What happens here is that there are actually two accretion processes, whose effects on the H II region must be carefully disentangled. The first accretion onto the sink particle has no effect on the H II region. Here, a blob of gas approaches the sink particle from behind, so that the continuum flux is not affected at all. Then a second blob crosses the line of sight, but it is still outside the accretion radius. Hence, it absorbs ionizing radiation, the H II region shrinks, but it is not yet accreted onto the sink particle. Only after it moves further towards the sink particle it gets inside the accretion radius. Thus, the important delay is the short gap between the shrinking of the H II region and the increase in the accretion rate.

Another difference between figure 4.26(a) and figure 4.26(b) is that the former only shows a singular accretion event while the latter mildly accretes over the whole interval. However, the evolution of H and $F_{2\,\mathrm{cm}}$ is not only governed by the material inside the accretion radius, so these curves do not exhibit a distinctively different behavior.

The analysis shows that when the accretion rate to the star has a large, sudden increase, the ionized region shrinks, and then slowly expands again. This agrees with the contraction, changes in shape, or anisotropic expansion observed in radio continuum observations of UC H II regions over intervals of $\sim 10\,\mathrm{yr}$ (Franco-Hernández & Rodríguez 2004, Rodríguez et al. 2007, Galván-Madrid et al. 2008, Gómez et al. 2008). Figure 4.26 shows that the sudden accretion of large amounts of material is

---

[2]The analysis presented here was carried out in collaboration with Roberto Galván-Madrid.

4 Results

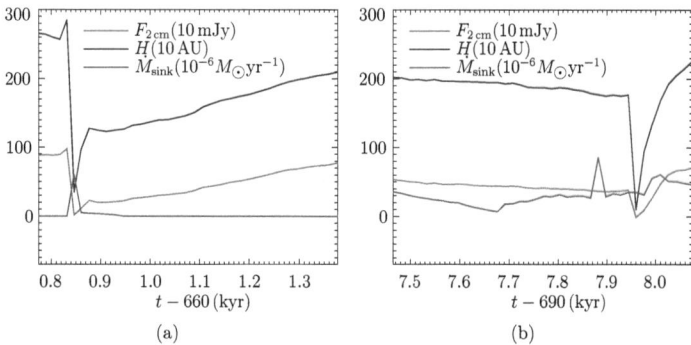

Figure 4.26: Time variability of UC H II regions. Large amounts of molecular gas accreting onto an H II region can cause a sudden decrease in its size and flux. This figure shows the 2-cm continuum flux $F_{2\,\mathrm{cm}}$ (in units of 10 mJy), the characteristic size of the H II region H (in units of 10 AU), and the rate of accretion to the star $\dot{M}$ (in units of $10^{-6}\,M_\odot\,\mathrm{yr}^{-1}$). (a) The H II region is initially relatively large, and accretion is almost shut off. A large (from 0 to $6 \cdot 10^{-5}\,M_\odot\,\mathrm{yr}^{-1}$), sudden accretion event causes the H II region to shrink and decrease in flux. The star at this moment has a mass of $19.8 M_\odot$. (b) The star has a larger mass ($23.3 M_\odot$), the H II region is initially smaller, and the star is constantly accreting gas. The ionizing photon flux appears to be able to ionize the infalling gas stably, until a peak in the accretion rate by a factor of three and the subsequent continuos accretion of gas makes the H II region to shrink and decrease in flux. The H II region does not shrink immediately after the accretion peak because the increase is relatively mild and the geometry of the infalling gas permitted ionizing photons to escape in one direction.

accompanied by a fast decrease in the observed size and flux of the H II region. Both the scale length and flux decrease at rates of 5–7 % yr^{-1}, agreeing well with observed fluctuations of 2–9 % per year (Franco-Hernández & Rodríguez 2004, Galván-Madrid et al. 2008). Shortly after the minimum values are reached, the H II regions starts growing anew, on timescales $\sim 10^2$ yr. This demonstrates that the observed time variability of UC H II regions is closely related to the interaction of the ionizing radiation of the massive protostar with its accretion flow.

5 Conclusions and Outlook

> So eine Arbeit wird eigentlich nie fertig, man muß sie für fertig erklären, wenn man nach Zeit und Umständen das möglichste getan hat.
>
> *(Johann Wolfgang von Goethe, Italienische Reise)*

We have presented the first three-dimensional collapse simulations of massive star formation that include radiative feedback by both ionizing and non-ionizing radiation. Our simulations show that the many puzzles raised by observations of UC H II regions, including the lifetime paradox, H II region morphologies and multiplicities, short-term variations in H II region flux and size, and uncollimated bipolar outflows around massive protostars can all be understood as consequences of ionization within a gravitationally unstable accretion flow. Massive stars require just such flows to form.

Our results from section 4.3.1 indicate that feedback by ionizing radiation does not represent a problem for massive star formation. If enough gas is available, it always finds its way onto the protostar. This is because massive stars require high accretion rates to form, high accretion rates correspond to massive accretion flows, and these accretion flows are subject to fragmentation. The fragments form filaments which then shield the ionizing radiation and allow accretion to continue. As laid out in section 4.3.2, the fragments in the accretion flow around a massive star can themselves form additional stars, which explains why massive stars preferentially form in clusters. If multiple star formation is taken into account, the upper masses of the stars are limited by fragmentation-induced starvation. This mechanism has not been discussed so far in the literature. It represents a new paradigm, different from monolithic collapse and competitive accretion. Since we only conducted simulations with a single initial condition, this gives rise to a couple of questions concerning the importance of this initial condition.

What determines the maximum mass in the cluster? Does the maximum mass scale with the initial gas reservoir, or does fragmentation constitute an unsurmountable barrier? If so, how do the extremely massive stars in the Galaxy form? Can the maximum mass be scaled up by choosing different, probably more centrally concentrated, initial conditions? Which role do magnetic fields play? How do the results change when turbulence is added to the initial core or when a box with decaying turbulence is considered? Do the clumps found in simulations of molecular cloud formation (Banerjee et al. 2009) behave similar like quiescent or turbulent cores, or do they represent another class of initial conditions? As explained in section 1.2, the choice of initial conditions is crucial to the formation of massive stars, and no definitive answer will be available until we know more about molecular cloud formation.

5 Conclusions and Outlook

The unique feature of massive star formation is that massive stars can ionize their own accretion flow. Section 4.3.4 has shown how we can understand observations of such flows, which show bipolar outflows, spiraling ionized gas and H II regions offset from the massive protostars. The numerical simulation is not only in agreement with an earlier analytical model (Keto 2007), but can also reproduce the actual observations (Zhang et al. 1998, Keto & Klaassen 2008), although it was not designed or tuned to do so. This suggests that the general mechanisms presented here, the fragmentation of the accretion flow (section 4.3.3) and the formation of ionization-driven bipolar outflows (section 4.3.4), are not special to the chosen initial conditions, but generic features of massive star formation.

The synthetic radio continuum maps (section 4.3.5) show remarkable similarity to morphologies found in surveys of massive star forming regions (Wood & Churchwell 1989, Kurtz et al. 1994). All morphologies can be found in the simulations. Their variety results solely from the interaction of the ionizing radiation with the accretion flow. We need neither bow shocks, stellar winds nor any other additional, special circumstances to reproduce them. Additionally, the embedding of the massive protostar in an accretion flow causes flickering of the H II region on a short timescale (setion 4.3.6) and renders any direct relation between the age of the protostar and the size of its H II region impossible. This resolves the lifetime problem for UC H II regions (Wood & Churchwell 1989) and explains observations of shrinking H II regions within only a decade (Franco-Hernández & Rodríguez 2004, Rodríguez et al. 2007, Galván-Madrid et al. 2008, Gómez et al. 2008).

All these observations demand follow-up studies. In which sense are the ionization-driven outflows different from magnetically-driven outflows found in low-mass star formation? Can these collimated, magnetically-driven jets survive inside the larger, uncollimated ionization-driven outflow, or is it destroyed by the radiative feedback? Can the interplay between ionizing radiation and magnetic fields generate new mechanisms to drive bipolar outflows?

The radio continuum maps of the UC H II regions are already very revealing, but they only trace the location of ionized gas, not its velocity. Can synthetic radio recombination line maps also reproduce the observed kinematics of UC H II regions (Jaffe et al. 2003, Zhu et al. 2008)? What do the morphologies look like at different wavelengths than those observed by the VLA? What do the simulations predict for upcoming telescopes like ALMA and JWST? What does the dust emission look like? In particular, is dust relevant to explain the abnormal spectral indices found in hypercompact H II regions, or are they caused by density gradients alone (Lizano 2008)?

Both the simulations and the postprocessing can be improved. Additional feedback effects like radiation pressure, stellar winds and outflows can be included. The simulation data can then be used to produce detailed synthetic observations that can be tested against reality. This will help us to decipher the mysteries of massive star formation.

Bibliography

O. Agertz, B. Moore, J. Stadel, D. Potter, F. Miniati, J. Read, L. Mayer, A. Gawryszczak, A. Kravtsov, Å. Nordlund, F. Pearce, V. Quilis, D. Rudd, V. Springel, J. Stone, E. Tasker, R. Teyssier, J. Wadsley and R. Walder. Fundamental differences between SPH and grid methods. *Monthly Notices of the Royal Astronomical Society*, vol. 380, pp. 963–978, 2007.

J. Alves, M. Lombardi and C. J. Lada. The mass function of dense molecular cores and the origin of the IMF. *Astronomy & Astrophysics*, vol. 462, pp. L17–L21, 2007.

J. F. Alves, C. J. Lada and E. A. Lada. Internal structure of a cold dark molecular cloud inferred from the extinction of background starlight. *Nature*, vol. 409, pp. 159–161, 2001.

P. André, D. Ward-Thompson and M. Barsony. From Prestellar Cores to Protostars: the Initial Conditions of Star Formation. In V. Mannings, A. P. Boss and S. S. Russell, editors, *Protostars and Planets IV*, pages 59–96. The University of Arizona Press, 2000.

J. Arons. Photon bubbles: Overstability in a magnetized atmosphere. *The Astrophysical Journal*, vol. 388, pp. 561–578, 1992.

S. J. Arthur and M. G. Hoare. Hydrodynamics of Cometary Compact H II Regions. *The Astrophysical Journal Supplement Series*, vol. 165, pp. 283–306, 2006.

M. Avalos, S. Lizano, L. F. Rodríguez, R. Franco-Hernández and J. M. Moran. Spectra and Sizes of Hypercompact H II Regions. *The Astrophysical Journal*, vol. 641, pp. 406–409, 2006.

R. Banerjee and R. E. Pudritz. Massive Star Formation via High Accretion Rates and Early Disk-driven Outflows. *The Astrophysical Journal*, vol. 660, pp. 479–488, 2007.

R. Banerjee, R. E. Pudritz and L. Holmes. The formation and evolution of protostellar discs; three-dimensional adaptive mesh refinement hydrosimulations of collapsing, rotating Bonnor-Ebert spheres. *Monthly Notices of the Royal Astronomical Society*, vol. 355, pp. 248–272, 2004.

R. Banerjee, R. E. Pudritz and D. W. Anderson. Supersonic turbulence, filamentary accretion and the rapid assembly of massive stars and discs. *Monthly Notices of the Royal Astronomical Society*, vol. 373, pp. 1091–1106, 2006.

R. Banerjee, R. S. Klessen and C. Fendt. Can Protostellar Jets Drive Supersonic Turbulence in Molecular Clouds? *The Astrophysical Journal*, vol. 668, pp. 1028–1041, 2007.

R. Banerjee, E. Vázquez-Semadeni, P. Hennebelle and R. S. Klessen. Clump morphology and evolution in MHD simulations of molecular cloud formation. *Monthly Notices of the Royal Astronomical Society*, vol. 398, pp. 1082–1092, 2009.

A. Barniske, L. M. Oskinova and W.-R. Hamann. Two extremely luminous WN stars in the Galactic center with circumstellar emission from dust and gas. *Astronomy & Astrophysics*, vol. 486, pp. 971–984, 2008.

M. R. Bate. Predicting the properties of binary stellar systems: the evolution of accreting protobinary systems. *Monthly Notices of the Royal Astronomical Society*, vol. 314, pp. 33–53, 2002.

M. R. Bate, I. A. Bonnell and N. M. Price. Modelling accretion in protobinary systems. *Monthly Notices of the Royal Astronomical Society*, vol. 277, pp. 363–376, 1995.

M. R. Bate, I. A. Bonnell and V. Bromm. The formation of a star cluster: predicting the properties of stars and brown dwarfs. *Monthly Notices of the Royal Astronomical Society*, vol. 339, pp. 577–599, 2003.

M. T. Beltrán, R. Cesaroni, C. Codella, L. Testi, R. S. Furuya and L. Olmi. Infall of gas as the formation mechanism of stars up to 20 times more massive than the Sun. *Nature*, vol. 443, pp. 427–429, 2006.

M. J. Berger and P. Colella. Local Adaptive Mesh Refinement for Shock Hydrodynamics. *Journal of Computational Physics*, vol. 82, pp. 64–84, 1989.

M. J. Berger and J. Oliger. Adaptive Mesh Refinement for Hyperbolic Partial Differential Equations. *Journal of Computational Physics*, vol. 53, pp. 484–512, 1984.

H. Beuther and P. Schilke. Fragmentation in Massive Star Formation. *Science*, vol. 303, pp. 1167–1169, 2004.

H. Beuther, Q. Zhang, L. J. Greenhill, M. J. Reid, D. Wilner, E. Keto, D. Marrone, P. T. P. Ho, J. M. Moran, R. Rao, H. Shinnaga and S.-Y. Liu. Subarcsecond Submillimeter Continuum Observations of Orion KL. *The Astrophysical Journal*, vol. 616, pp. L31–L34, 2004.

J. Binney and S. Tremaine. *Galactic Dynamics*. Princeton University Press, second edition, 2008.

T. G. Bisbas, R. Wünsch, A. P. Whitworth and D. A. Hubber. Smoothed particle hydrodynamics simulations of expanding H II regions. I. Numerical method and applications. *Astronomy & Astrophysics*, vol. 497, pp. 649–659, 2009.

I. A. Bonnell, M. R. Bate, C. J. Clarke and J. E. Pringle. Accretion and the stellar mass spectrum in small clusters. *Monthly Notices of the Royal Astronomical Society*, vol. 285, pp. 201–208, 1997.

I. A. Bonnell, M. R. Bate and H. Zinnecker. On the formation of massive stars. *Monthly Notices of the Royal Astronomical Society*, vol. 298, pp. 93–102, 1998.

I. A. Bonnell, M. R. Bate, C. J. Clarke and J. E. Pringle. Competitive accretion in embedded stellar clusters. *Monthly Notices of the Royal Astronomical Society*, vol. 323, pp. 785–794, 2001.

I. A. Bonnell, S. G. Vine and M. R. Bate. Massive star formation: nurture, not nature. *Monthly Notices of the Royal Astronomical Society*, vol. 349, pp. 735–741, 2004.

I. A. Bonnell, R. B. Larson and H. Zinnecker. The Origin of the Initial Mass Function. In B. Reipurth, D. Jewitt and K. Keil, editors, *Protostars and Planets V*, pages 149–164. The University of Arizona Press, 2007.

W. B. Bonnor. Boyle's Law and gravitational instability. *Monthly Notices of the Royal Astronomical Society*, vol. 116, pp. 351–359, 1956.

A. P. Boss and P. Bodenheimer. Fragmentation in a rotating protostar: A comparison of two three-dimensional computer codes. *The Astrophysical Journal*, vol. 234, pp. 289–295, 1979.

A. C. Calder, B. C. Curtis, L. J. Dursi, B. Fryxell, G. Henry, P. MacNeice, K. Olson, P. Ricker, R. Rosner, F. X. Timmes, H. M. Tufo, J. W. Truran and M. Zingale. High performance reactive fluid flow simulations using adaptive mesh refinement on thousands of processors. In *Supercomputing '00: Proceedings of the 2000 ACM/IEEE conference on Supercomputing (CDROM)*, Washington, DC, USA, 2000. IEEE Computer Society.

A. C. Calder, B. Fryxell, T. Plewa, R. Rosner, L. J. Dursi, V. G. Weirs, T. Dupont, H. F. Robey, J. O. Kane, B. A. Remington, R. P. Drake, G. Dimonte, M. Zingale, F. X. Timmes, K. Olson, P. Ricker, P. MacNeice and H. M. Tufo. On Validating an Astrophysical Simulation Code. *The Astrophysical Journal Supplement Series*, vol. 143, pp. 201–229, 2002.

J. Castor. *Radiation Hydrodynamics*. Cambridge University Press, 2004.

G. Chabrier. Galactic Stellar and Substellar Initial Mass Function. *Publications of the Astronomical Society of the Pacific*, vol. 115, pp. 763–795, 2003.

R. Chini, V. Hoffmeister, S. Kimeswenger, M. Nielbock, D. Nürnberger, L. Schmidtobreick and M. Sterzik. The formation of a massive protostar through the disk accretion of gas. *Nature*, vol. 429, pp. 155–157, 2004.

A. J. Chorin and J. E. Marsden. *A Mathematical Introduction to Fluid Mechanics*. Springer-Verlag, third edition, 2000.

K.-H. W. Chu. A pedagogical look at Jeans' density scale. *European Journal of Physics*, vol. 28, pp. 501–507, 2007.

E. Churchwell. Ultra-Compact HII Regions and Massive Star Formation. *Annual Reviews of Astronomy & Astrophysics*, vol. 40, pp. 27–62, 2002.

P. C. Clark, I. A. Bonnell and R. S. Klessen. The star formation efficiency and its relation to variations in the initial mass function. *Monthly Notices of the Royal Astronomical Society*, vol. 386, pp. 3–10, 2008.

C. J. Clarke and R. F. Carswell. *Principles of Astrophysical Fluid Dynamics*. Cambridge University Press, 2007.

P. Colella and H. M. Glaz. Efficient Solution Algorithms for the Riemann Problem for Real Gases. *Journal of Computational Physics*, vol. 59, pp. 264–289, 1985.

P. Colella and P. R. Woodward. The Piecewise Parabolic Method (PPM) for Gas-Dynamical Simulations. *Journal of Computational Physics*, vol. 54, pp. 174–201, 1984.

B. Commerçon, P. Hennebelle, E. Audit, G. Chabrier and R. Teyssier. Protostellar collapse: a comparison between smoothed particle hydrodynamics and adaptative mesh refinement calculations. *Astronomy & Astrophysics*, vol. 482, pp. 371–385, 2008.

R. Courant, K. Friedrichs and H. Lewy. Über die partiellen Differenzengleichungen der mathematischen Physik. *Mathematische Annalen*, vol. 100, pp. 32–74, 1928.

D. P. Cox and W. H. Tucker. Ionization Equilibrium and Radiative Cooling of a Low-Density Plasma. *The Astrophysical Journal*, vol. 157, pp. 1157–1167, 1969.

J. E. Dale, I. A. Bonnell, C. J. Clarke and M. R. Bate. Photoionizing feedback in star cluster formation. *Monthly Notices of the Royal Astronomical Society*, vol. 358, pp. 291–304, 2005.

J. E. Dale, I. A. Bonnell and A. P. Whitworth. Ionization-induced star formation — I. The collect-and-collapse model. *Monthly Notices of the Royal Astronomical Society*, vol. 375, pp. 1291–1298, 2007.

J. E. Dale, P. C. Clark and I. A. Bonnell. Ionization-induced star formation — II. External irradiation of a turbulent molecular cloud. *Monthly Notices of the Royal Astronomical Society*, vol. 377, pp. 534–544, 2007.

A. Dalgarno and R. A. McCray. Heating and Ionization of HI Regions. *Annual Reviews of Astronomy & Astrophysics*, vol. 10, pp. 375–426, 1972.

Bibliography

P. A. Davidson. *Turbulence*. Oxford University Press, 2004.

C. G. De Pree, L. F. Rodríguez and W. M. Goss. Ultracompact H II regions: are their lifetimes extended by dense, warm environments? *Revista Mexicana de Astronomía y Astrofísica*, vol. 31, pp. 39–44, 1995.

C. G. De Pree, D. J. Wilner, J. Deblasio, A. J. Mercer and L. E. Davis. The Morphologies of Ultracompact H II Regions in W49A and Sagittarius B2: The Prevalence of Shells and a Modified Classification Scheme. *The Astrophysical Journal*, vol. 624, pp. L101–L104, 2005.

D. De Zeeuw and K. G. Powell. An Adaptively Refined Cartesian Mesh Solver for the Euler Equations. *Journal of Computational Physics*, vol. 104, pp. 56–68, 1993.

C. L. Dobbs, I. A. Bonnell and P. C. Clark. Centrally condensed turbulent cores: massive stars or fragmentation? *Monthly Notices of the Royal Astronomical Society*, vol. 360, pp. 2–8, 2005.

J. E. Dyson and D. A. Williams. *Physics of the interstellar medium*. Manchester University Press, 1980.

R. Ebert. Über die Verdichtung von H I-Gebieten. *Zeitschrift für Astrophysik*, vol. 37, pp. 217–232, 1955.

R. Edgar and C. Clarke. The effect of radiative feedback on Bondi-Hoyle flow around a massive star. *Monthly Notices of the Royal Astronomical Society*, vol. 349, pp. 678–686, 2004.

B. G. Elmegreen and C. J. Lada. Sequential formation of subgroups in OB associations. *The Astrophysical Journal*, vol. 214, pp. 725–741, 1977.

B. G. Elmegreen and J. Scalo. Interstellar Turbulence I: Observations and Processes. *Annual Reviews of Astronomy & Astrophysics*, vol. 42, pp. 211–273, 2004.

A. Esquivel and A. C. Raga. Radiation-driven collapse of autogravitating neutral clumps. *Monthly Notices of the Royal Astronomical Society*, vol. 377, pp. 383–390, 2007.

N. J. Evans, II. Physical Conditions in Regions of Star Formation. *Annual Reviews of Astronomy & Astrophysics*, vol. 37, pp. 311–362, 1999.

C. Federrath, P. C. Clark, R. Banerjee and R. S. Klessen. Implementation of Sink Particles in the FLASH Code. *in preparation*, 2009.

J. W. Ferguson, D. R. Alexander, F. Allard, T. Barman, J. G. Bodnarik, P. H. Hauschildt, A. Heffner-Wong and A. Tamanai. Low-Temperature Opacities. *The Astrophysical Journal*, vol. 623, pp. 585–596, 2005.

D. F. Figer. An upper limit to the masses of stars. *Nature*, vol. 434, pp. 192–194, 2005.

P. N. Foster and R. A. Chevalier. Gravitational Collapse of an Isothermal Sphere. *The Astrophysical Journal*, vol. 416, pp. 303–311, 1993.

J. Franco, G. Tenorio-Tagle and P. Bodenheimer. On the formation and expansion of H II regions. *The Astrophysical Journal*, vol. 349, pp. 126–140, 1990.

J. Franco, S. Kurtz, P. Hofner, L. Testi, G. García-Segura and M. Martos. The Density Structure of Highly Compact H II Regions. *The Astrophysical Journal*, vol. 542, pp. L143–L146, 2000.

R. Franco-Hernández and L. F. Rodríguez. Time Variation in the Radio Flux Density from the Bipolar Ultracompact H II Region NGC 7538 IRS 1. *The Astrophysical Journal*, vol. 604, pp. L105–L108, 2004.

A. Frank and G. Mellema. A radiation-gasdynamical method for numerical simulations of ionized nebulae: Radiation-gasdynamics of PNe I. *Astronomy & Astrophysics*, vol. 289, pp. 937–945, 1994.

J. Frank, A. King and D. Raine. *Accretion power in astrophysics*. Cambridge University Press, second edition, 1992.

T. Freyer, G. Hensler and H. W. Yorke. Massive Stars and the Energy Balance of the Interstellar Medium. I. The Impact of an Isolated 60 $M_\odot$ Star. *The Astrophysical Journal*, vol. 594, pp. 888–910, 2003.

T. Freyer, G. Hensler and H. W. Yorke. Massive Stars and the Energy Balance of the Interstellar Medium. II. The 35 $M_\odot$ Star and a Solution to the "Missing Wind Problem". *The Astrophysical Journal*, vol. 638, pp. 262–280, 2006.

B. Fryxell, K. Olson, P. Ricker, F. X. Timmes, M. Zingale, D. Q. Lamb, P. MacNeice, R. Rosner, J. W. Truran and H. Tufo. FLASH: An Adaptive Mesh Hydrodynamics Code for Modeling Astrophysical Thermonuclear Flashes. *The Astrophysical Journal Supplement Series*, vol. 131, pp. 273–334, 2000.

R. Galván-Madrid, L. F. Rodríguez, P. T. P. Ho and E. Keto. Time Variation in G24.78+0.08 A1: Evidence for an Accreting Hypercompact H II Region? *The Astrophysical Journal*, vol. 674, pp. L33–L36, 2008.

C. F. Gammie. Photon bubbles in accretion disks. *Monthly Notices of the Royal Astronomical Society*, vol. 297, pp. 929–935, 1998.

G. Garay, L. F. Rodríguez, J. M. Moran and E. Churchwell. VLA Observations of Strong IRAS Point Sources Associated with Compact H II Regions. *The Astrophysical Journal*, vol. 418, pp. 368–385, 1993.

G. García-Segura and J. Franco. From Ultracompact to Extended H II Regions. *The Astrophysical Journal*, vol. 469, pp. 171–188, 1996.

G. García-Segura, J. A. López, W. Steffen, J. Meaburn and A. Manchado. The Dynamical Evolution of Planetary Nebulae after the Fast Wind. *The Astrophysical Journal*, vol. 646, pp. L61–L64, 2006.

R. A. Gingold and J. J. Monaghan. Smoothed particle hydrodynamics: theory and application to non-spherical stars. *Monthly Notices of the Royal Astronomical Society*, vol. 181, pp. 375–389, 1977.

J. L. Giuliani, Jr. The hydrodynamic stability of ionization-shock fronts. Linear theory. *The Astrophysical Journal*, vol. 233, pp. 280–293, 1979.

S. C. O. Glover and M.-M. Mac Low. Simulating the Formation of Molecular Clouds. I. Slow Formation by Gravitational Collapse from Static Initial Conditions. *The Astrophysical Journal Supplement Series*, vol. 169, pp. 239–268, 2007.

P. F. Goldsmith. Molecular Depletion and Thermal Balance in Dark Cloud Cores. *The Astrophysical Journal*, vol. 557, pp. 736–746, 2001.

L. Gómez, L. F. Rodríguez, L. Loinard, S. Lizano, C. Allen, A. Poveda and K. M. Menten. Monitoring the Large Proper Motions of Radio Sources in the Orion BN/KL Region. *The Astrophysical Journal*, vol. 685, pp. 333–343, 2008.

M. A. Gordon and R. L. Sorochenko. *Radio Recombination Lines*. Kluwer Academic Publishers Group, 2002.

M. Gritschneder, T. Naab, A. Burkert, S. Walch, F. Heitsch and M. Wetzstein. iVINE — Ionization in the parallel TREE/SPH code VINE: first results on the observed age-spread around O-stars. *Monthly Notices of the Royal Astronomical Society*, vol. 393, pp. 21–31, 2009.

M. Gritschneder, T. Naab, S. Walch, A. Burkert and F. Heitsch. Driving Turbulence and Triggering Star Formation by Ionizing Radiation. *The Astrophysical Journal*, vol. 694, pp. L26–L30, 2009.

W. Hackbusch. *Multi-Grid Methods and Applications*, vol. 4 of *Springer Series in Computational Mathematics*. Springer-Verlag, 1985.

G. H. Herbig. The properties and problems of T Tauri stars and related objects. *Advances in Astronomy and Astrophysics*, vol. 1, pp. 47–103, 1962.

J. J. Hester and S. J. Desch. Understanding Our Origins: Star Formation in HII Region Environments. In A. N. Krot, E. R. D. Scott and B. Reipurth, editors, *Chondrites and the Protoplanetary Disk*, vol. 341 of *Astronomical Society of the Pacific Conference Series*, pages 107–130, 2005.

P. T. P. Ho and A. D. Haschick. Formation of OB clusters: VLA observations. *The Astrophysical Journal*, vol. 248, pp. 622–637, 1981.

P. T. P. Ho and C. H. Townes. Interstellar ammonia. *Annual Reviews of Astronomy & Astrophysics*, vol. 21, pp. 239–270, 1983.

D. Hollenbach and C. F. McKee. Molecule formation and infrared emission in fast interstellar shocks. I. Physical processes. *The Astrophysical Journal Supplement Series*, vol. 41, pp. 555–592, 1979.

D. Hollenbach, D. Johnstone, S. Lizano and F. Shu. Photoevaporation of disks around massive stars and application to ultracompact H II regions. *The Astrophysical Journal*, vol. 428, pp. 654–669, 1994.

T. Hosokawa and K. Omukai. Evolution of Massive Protostars with High Accretion Rates. *The Astrophysical Journal*, vol. 691, pp. 823–846, 2009.

R. Q. Huang and K. N. Yu. *Stellar Astrophysics*. Springer-Verlag, 1998.

C. Hunter. The Instability of the Collapse of a Self-Gravitating Gas Cloud. *The Astrophysical Journal*, vol. 136, pp. 594–608, 1962.

C. Hunter. The collapse of unstable isothermal spheres. *The Astrophysical Journal*, vol. 218, pp. 834–845, 1977.

I. Iben, Jr. Stellar Evolution. I. The Approach to the Main Sequence. *The Astrophysical Journal*, vol. 141, pp. 993–1018, 1965.

R. Ignace and E. Churchwell. Free-Free Spectral Energy Distributions of Hierarchically Clumped H II Regions. *The Astrophysical Journal*, vol. 610, pp. 351–360, 2004.

I. T. Iliev, B. Ciardi, M. A. Alvarez, A. Maselli, A. Ferrara, N. Y. Gnedin, G. Mellema, T. Nakamoto, M. L. Norman, A. O. Razoumov, E.-J. Rijkhorst, J. Ritzerveld, P. R. Shapiro, H. Susa, M. Umemura and D. J. Whalen. Cosmological radiative transfer codes comparison project - I. The static density field tests. *Monthly Notices of the Royal Astronomical Society*, vol. 371, pp. 1057–1086, 2006.

I. T. Iliev, D. Whalen, G. Mellema, K. Ahn, S. Baek, N. Y. Gnedin, A. V. Kravtsov, M. Norman, M. Raicevic, D. R. Reynolds, D. Sato, P. R. Shapiro, B. Semelin, J. Smidt, H. Susa, T. Theuns and M. Umemura. Cosmological Radiative Transfer Comparison Project II: The Radiation-Hydrodynamic Tests. *arxiv:0905.2920v1*, 2009.

D. T. Jaffe, Q. Zhu, J. H. Lacy and M. Richter. Kinematics of Ultracompact H II Regions Revealed: High Spatial and Spectral Resolution Mapping of the 12.8 Micron [Ne II] Line in Monoceros R2. *The Astrophysical Journal*, vol. 596, pp. 1053–1063, 2003.

R. A. James. The Solution of Poisson's Equation for Isolated Source Distributions. *Journal of Computational Physics*, vol. 25, pp. 71–93, 1977.

J. H. Jeans. The Stability of a Spherical Nebula. *Philosophical Transactions of the Royal Society of London. Series A*, vol. 199, pp. 1–53, 1902.

J. Jijina and F. C. Adams. Infall Collapse Solutions in the Inner Limit: Radiation Pressure and Its Effects on Star Formation. *The Astrophysical Journal*, vol. 462, pp. 874–887, 1996.

F. D. Kahn. Cocoons around Early-type Stars. *Astronomy & Astrophysics*, vol. 37, pp. 149–162, 1974.

O. Kessel-Deynet and A. Burkert. Radiation-driven implosion of molecular cloud cores. *Monthly Notices of the Royal Astronomical Society*, vol. 338, pp. 545–554, 2003.

E. Keto. An Ionized Accretion Flow in the Ultracompact H II Region G10.6-0.4. *The Astrophysical Journal*, vol. 568, pp. 754–760, 2002.

E. Keto. On the Evolution of Ultracompact H II Regions. *The Astrophysical Journal*, vol. 580, pp. 980–986, 2002.

E. Keto. The Formation of Massive Stars: Accretion, Disks, and the Development of Hypercompact H II Regions. *The Astrophysical Journal*, vol. 666, pp. 976–981, 2007.

E. Keto and P. Klaassen. The Ionization of Accretion Flows in High-Mass Star Formation: W51e2. *The Astrophysical Journal*, vol. 678, pp. L109–L112, 2008.

E. Keto and K. Wood. Observations on the Formation of Massive Stars by Accretion. *The Astrophysical Journal*, vol. 637, pp. 850–859, 2006.

E. Keto, Q. Zhang and S. Kurtz. The Early Evolution of Massive Stars: Radio Recombination Line Spectra. *The Astrophysical Journal*, vol. 672, pp. 423–432, 2008.

E. R. Keto. Radiative transfer modeling of radio-frequency spectral line data: Accretion onto G10.6-0.4. *The Astrophysical Journal*, vol. 355, pp. 190–196, 1990.

E. R. Keto, P. T. P. Ho and A. D. Haschick. Temperature and density structure of the collapsing core of G10.6-0.4. *The Astrophysical Journal*, vol. 318, pp. 712–728, 1987.

M. K.-H. Kiessling. The "Jeans swindle": A true story—mathematically speaking. *Advances in Applied Mathematics*, vol. 31, pp. 132–149, 2003.

K.-T. Kim and B.-C. Koo. Radio Continuum and Recombination Line Study of Ultracompact H II Regions with Extended Envelopes. *The Astrophysical Journal*, vol. 549, pp. 979–996, 2001.

R. I. Klein, C. F. McKee and P. Colella. On the Hydrodynamic Interaction of Shock Waves with Interstellar Clouds. I. Nonradiative Shocks in Small Clouds. *The Astrophysical Journal*, vol. 420, pp. 213–236, 1994.

R. S. Klessen. *The Relation between Interstellar Turbulence and Star Formation.* Habilitationsschrift, Universität Potsdam, 2003.

R. S. Klessen and A. Burkert. The Formation of Stellar Clusters: Gaussian Cloud Conditions. I. *The Astrophysical Journal Supplement Series*, vol. 128, pp. 287–319, 2000.

R. S. Klessen, A. Burkert and M. R. Bate. Fragmentation of Molecular Clouds: The Initial Phase of a Stellar Cluster. *The Astrophysical Journal*, vol. 501, pp. L205–L208, 1998.

R. S. Klessen, M. R. Krumholz and F. Heitsch. Numerical Star-Formation Studies — A Status Report. *arxiv:0906.4452v1*, 2009.

K. M. Kratter and C. D. Matzner. Fragmentation of massive protostellar discs. *Monthly Notices of the Royal Astronomical Society*, vol. 373, pp. 1563–1576, 2006.

J. D. Kraus. *Radio Astronomy*. McGraw-Hill, Inc., 1966.

P. Kroupa. The Initial Mass Function of Stars: Evidence for Uniformity in Variable Systems. *Science*, vol. 295, pp. 82–91, 2002.

M. R. Krumholz. Radiation Feedback and Fragmentation in Massive Protostellar Cores. *The Astrophysical Journal*, vol. 641, pp. L45–L48, 2006.

M. R. Krumholz and I. A. Bonnell. Models for the Formation of Massive Stars. *arxiv: 0712.0828v2*, 2007.

M. R. Krumholz and C. D. Matzner. The Dynamics of Radiation-pressure-dominated H II Regions. *The Astrophysical Journal*, vol. 703, pp. 1352–1362, 2009.

M. R. Krumholz, C. F. McKee and R. I. Klein. Embedding Lagrangian Sink Particles in Eulerian Grids. *The Astrophysical Journal*, vol. 611, pp. 399–412, 2004.

M. R. Krumholz, R. I. Klein and C. F. McKee. Radiation pressure in massive star formation. In R. Cesaroni, M. Felli, E. Churchwell and M. Walmsley, editors, *Massive star birth: A crossroads of Astrophysics*, number 227 in IAU Symposium, pages 231–236. Cambridge University Press, 2005.

M. R. Krumholz, C. F. McKee and R. I. Klein. How Protostellar Outflows Help Massive Stars Form. *The Astrophysical Journal*, vol. 618, pp. L33–L36, 2005.

M. R. Krumholz, R. I. Klein and C. F. McKee. Radiation-Hydrodynamic Simulations of Collapse and Fragmentation in Massive Protostellar Cores. *The Astrophysical Journal*, vol. 656, pp. 959–979, 2007.

M. R. Krumholz, R. I. Klein and C. F. McKee. Equations and Algorithms for Mixed-frame Flux-limited Diffusion Radiation Hydrodynamics. *The Astrophysical Journal*, vol. 667, pp. 626–643, 2007.

M. R. Krumholz, J. M. Stone and T. A. Gardiner. Magnetohydrodynamic Evolution of H II Regions in Molecular Clouds: Simulation Methodology, Tests, and Uniform Media. *The Astrophysical Journal*, vol. 671, pp. 518–535, 2007.

M. R. Krumholz, R. I. Klein, C. F. McKee, S. S. R. Offner and A. J. Cunningham. The Formation of Massive Star Systems by Accretion. *Science*, vol. 323, pp. 754–757, 2009.

S. Kurtz. Hypercompact HII Regions. In R. Cesaroni, M. Felli, E. Churchwell and M. Walmsley, editors, *Massive star birth: A crossroads of Astrophysics*, number 227 in IAU Symposium, pages 111–119. Cambridge University Press, 2005.

S. Kurtz, E. Churchwell and D. O. S. Wood. Ultracompact H II regions. II. New high-resolution radio images. *The Astrophysical Journal Supplement Series*, vol. 91, pp. 659–712, 1994.

C. J. Lada and E. A. Lada. Embedded Clusters in Molecular Clouds. *Annual Reviews of Astronomy & Astrophysics*, vol. 41, pp. 57–115, 2003.

C. J. Lada, A. A. Muench, J. Rathborne, J. F. Alves and M. Lombardi. The Nature of the Dense Core Population in the Pipe Nebula: Thermal Cores Under Pressure. *The Astrophysical Journal*, vol. 672, pp. 410–422, 2008.

L. D. Landau and E. M. Lifschitz. *Hydrodynamik*. Akademie Verlag GmbH, fifth edition, 1991.

C. B. Laney. *Computational Gasdynamics*. Cambridge University Press, 1998.

R. B. Larson. The physics of star formation. *Reports on Progress in Physics*, vol. 66, pp. 1651–1697, 2003.

R. B. Larson. Insights from simulations of star formation. *Reports on Progress in Physics*, vol. 70, pp. 337–356, 2007.

R. B. Larson. Numerical calculations of the dynamics of a collapsing proto-star. *Monthly Notices of the Royal Astronomical Society*, vol. 145, pp. 271–295, 1969.

R. B. Larson and S. Starrfield. On the Formation of Massive Stars and the Upper Limit of Stellar Masses. *Astronomy & Astrophysics*, vol. 13, pp. 190–197, 1971.

S. Lepp and J. M. Shull. The kinetic theory of H_2 dissociation. *The Astrophysical Journal*, vol. 270, pp. 578–582, 1987.

R. J. LeVeque. *Numerical Methods for Conservation Laws*. Birkhäuser Verlag, second edition, 1994.

R. J. LeVeque, D. Mihalas, E. A. Dorfi and E. Müller. *Computational Methods for Astrophysical Fluid Flow*. Springer-Verlag, 1998.

S. Lizano. Hypercompact HII Regions. In H. Beuther, H. Linz and T. Henning, editors, *Massive Star Formation: Observations Confront Theory*, vol. 387 of *Astronomical Society of the Pacific Conference Series*, pages 232–239, 2008.

R. Löhner. An adaptive finite element scheme for transient problems in CFD. *Computer Methods in Applied Mechanics and Engineering*, vol. 61(3), pp. 323–338, 1987.

L. B. Lucy. A numerical approach to the testing of the fission hypothesis. *The Astronomical Journal*, vol. 82(12), pp. 1013–1024, 1977.

M.-M. Mac Low. Feedback Processes: A Theoretical Perspective. In H. Beuther, H. Linz and T. Henning, editors, *Massive Star Formation: Observations Confront Theory*, vol. 387 of *Astronomical Society of the Pacific Conference Series*, pages 148–157, 2008.

M.-M. Mac Low. The Energy Dissipation Rate of Supersonic, Magnetohydrodynamic Turbulence in Molecular Clouds. *The Astrophysical Journal*, vol. 524, pp. 169–178, 1999.

M.-M. Mac Low and R. S. Klessen. Control of star formation by supersonic turbulence. *Reviews of Modern Physics*, vol. 76, pp. 125–194, 2004.

M.-M. Mac Low and J. M. Shull. Molecular processes and gravitational collapse in intergalactic shocks. *The Astrophysical Journal*, vol. 302, pp. 585–589, 1986.

M.-M. Mac Low, D. Van Buren, D. O. S. Wood and E. Churchwell. Bow shock models of ultracompact H II regions. *The Astrophysical Journal*, vol. 369, pp. 395–409, 1991.

M.-M. Mac Low, J. Toraskar, J. S. Oishi and T. Abel. Dynamical Expansion of H II Regions from Ultracompact to Compact Sizes in Turbulent, Self-gravitating Molecular Clouds. *The Astrophysical Journal*, vol. 668, pp. 980–992, 2007.

P. MacNeice, K. M. Olson, C. Mobarry, R. de Fainchtein and C. Packer. PARAMESH: A parallel adaptive mesh refinement community toolkit. *Computer Physics Communications*, vol. 126, pp. 330–354, 2000.

J. E. Marsden and T. J. R. Hughes. *Mathematical foundations of elasticity*. Dover Publications, Inc., 1994.

D. F. Martin and K. L. Cartwright. Solving Poisson's Equation using Adaptive Mesh Refinement. Unpublished manuscript, available online from http://seesar.lbl.gov/anag/staff/martin/tar/AMR.ps, 1996.

P. G. Martin, W. J. Keogh and M. E. Mandy. Collision-induced Dissociation of Molecular Hydrogen at Low Densities. *The Astrophysical Journal*, vol. 499, pp. 793–798, 1998.

P. Massey. The Initial Mass Function of Massive Stars in the Local Group. In G. Gilmore and D. Howell, editors, *The Stellar Initial Mass Function*, vol. 142 of *Astronomical Society of the Pacific Conference Series*, pages 17–44, 1998.

C. F. McKee and J. C. Tan. Massive star formation in 100,000 years from turbulent and pressurized molecular clouds. *Nature*, vol. 416, pp. 59–61, 2002.

C. F. McKee and J. C. Tan. The Formation of Massive Stars from Turbulent Cores. *The Astrophysical Journal*, vol. 585, pp. 850–871, 2003.

G. Mellema and P. Lundqvist. Stellar wind bubbles around WR and [WR] stars. *Astronomy & Astrophysics*, vol. 394, pp. 901–909, 2002.

G. Mellema, S. J. Arthur, W. J. Henney, I. T. Iliev and P. R. Shapiro. Dynamical H II Region Evolution in Turbulent Molecular Clouds. *The Astrophysical Journal*, vol. 647, pp. 397–403, 2006.

J. Miao, G. J. White, R. Nelson, M. Thompson and L. Morgan. Triggered star formation in bright-rimmed clouds: the Eagle nebula revisited. *Monthly Notices of the Royal Astronomical Society*, vol. 369, pp. 143–155, 2006.

D. Mihalas and B. W. Mihalas. *Foundations of Radiation Hydrodynamics*. Oxford University Press, 1984.

J. J. Monaghan. Smoothed particle hydrodynamics. *Reports on Progress in Physics*, vol. 68, pp. 1703–1759, 2005.

J. J. Monaghan. Smoothed particle hydrodynamics. *Annual Reviews of Astronomy & Astrophysics*, vol. 30, pp. 543–574, 1992.

F. Motte, P. André and R. Neri. The initial conditions of star formation in the ρ Ophiuchi main cloud: wide-field millimeter continuum mapping. *Astronomy & Astrophysics*, vol. 336, pp. 150–172, 1998.

F. Motte, P. André, D. Ward-Thompson and S. Bontemps. A SCUBA survey of the NGC 2068/2071 protoclusters. *Astronomy & Astrophysics*, vol. 372, pp. L41–L44, 2001.

F. Motte, S. Bontemps, N. Schneider, P. Schilke and K. M. Menten. Massive Infrared-Quiet Dense Cores: Unveiling the Initial Conditions of High-Mass Star Formation. In H. Beuther, H. Linz and T. Henning, editors, *Massive Star Formation: Observations Confront Theory*, vol. 387 of *Astronomical Society of the Pacific Conference Series*, pages 22–29, 2008.

P. C. Myers, T. M. Dame, P. Thaddeus, R. S. Cohen, R. F. Silverberg, E. Dwek and M. G. Hauser. Molecular clouds and star formation in the inner galaxy: A comparison of CO, H II, and far-infrared surveys. *The Astrophysical Journal*, vol. 301, pp. 398–422, 1986.

F. Nakamura and Z.-Y. Li. Protostellar Turbulence Driven by Collimated Outflows. *The Astrophysical Journal*, vol. 662, pp. 395–412, 2007.

F. Nakamura, C. F. McKee, R. I. Klein and R. T. Fisher. On the Hydrodynamic Interaction of Shock Waves with Interstellar Clouds. II. The Effect of Smooth Cloud Boundaries on Cloud Destruction and Cloud Turbulence. *The Astrophysical Journal Supplement Series*, vol. 164, pp. 477–505, 2006.

T. Nakano. Conditions for the formation of massive stars through nonspherical accretion. *The Astrophysical Journal*, vol. 345, pp. 464–471, 1989.

T. Nakano, T. Hasegawa and C. Norman. The Mass of a Star Formed in a Cloud Core: Theory and Its Application to the Orion A Cloud. *The Astrophysical Journal*, vol. 450, pp. 183–195, 1995.

D. A. Neufeld, S. Lepp and G. J. Melnick. Thermal Balance in Dense Molecular Clouds: Radiative Cooling Rates and Emission-Line Luminosities. *The Astrophysical Journal Supplement Series*, vol. 100, pp. 132–147, 1995.

D. E. A. Nürnberger, R. Chini, F. Eisenhauer, M. Kissler-Patig, A. Modigliani, R. Siebenmorgen, M. F. Sterzik and T. Szeifert. Formation of a massive protostar through disk accretion. II. SINFONI integral field spectroscopy of the M 17 silhouette disk and discovery of the associated H_2 jet. *Astronomy & Astrophysics*, vol. 465, pp. 931–936, 2007.

F. M. Olnon. Thermal Bremsstrahlung Radiospectra for Inhomogeneous Objects, with an Application to MWC 349. *Astronomy & Astrophysics*, vol. 39, pp. 217–223, 1975.

J. H. Oort and L. Spitzer, Jr. Acceleration of Interstellar Clouds by O-Type Stars. *The Astrophysical Journal*, vol. 121, pp. 6–23, 1955.

V. Ossenkopf and M.-M. Mac Low. Turbulent velocity structure in molecular clouds. *Astronomy & Astrophysics*, vol. 390, pp. 307–326, 2002.

D. E. Osterbrock. *Astrophysics of Gaseous Nebulae and Active Galactic Nuclei*. University Science Books, 1989.

T. Padmanabhan. *Theoretical Astrophysics*, vol. I: Astrophysical Processes. Cambridge University Press, 2000.

F. Palla and S. W. Stahler. The Pre-Main-Sequence Evolution of Intermediate-Mass Stars. *The Astrophysical Journal*, vol. 418, pp. 414–425, 1993.

N. Panagia and M. Felli. The Spectrum of the Free-free Radiation from Extended Envelopes. *Astronomy & Astrophysics*, vol. 39, pp. 1–5, 1975.

B. Paxton. EZ to Evolve ZAMS Stars: A Program Derived from Eggleton's Stellar Evolution Code. *Publications of the Astronomical Society of the Pacific*, vol. 116, pp. 699–701, 2004.

M. V. Penston. Dynamics of self-gravitating gaseous spheres—III. Analytical results in the free-fall of isothermal cases. *Monthly Notices of the Royal Astronomical Society*, vol. 144, pp. 425–448, 1969.

A. Peraiah. *An Introduction to Radiative Transfer*. Cambridge University Press, 2002.

T. Peters, R. Banerjee and R. S. Klessen. Ionization front-driven turbulence in the clumpy interstellar medium. *Physica Scripta*, vol. T132, p. 014026, 2008.

J. B. Pollack, D. Hollenbach, S. Beckwith, D. P. Simonelli, T. Roush and W. Fong. Composition and radiative properties of grains in molecular clouds and accretion disks. *The Astrophysical Journal*, vol. 421, pp. 615–639, 1994.

P. Pratap, L. E. Snyder and W. Batrla. Are the NGC 7538 formaldehyde masers really unusual? *The Astrophysical Journal*, vol. 387, pp. 241–247, 1992.

W. H. Press, B. P. Flannery, S. A. Teukolsky and W. T. Vetterling. *Numerical Recipes*. Cambridge University Press, 1986.

D. J. Price. Modelling discontinuities and Kelvin-Helmholtz instabilities in SPH. *Journal of Computational Physics*, vol. 227, pp. 10040–10057, 2008.

R. E. Pudritz. Clustered Star Formation and the Origin of Stellar Masses. *Science*, vol. 295, pp. 68–76, 2002.

A. C. Raga, W. Henney, J. Vasconcelos, A. Cerqueira, A. Esquivel and A. Rodríguez-González. Multiple clump structures within photoionized regions. *Monthly Notices of the Royal Astronomical Society*, vol. 392, pp. 964–968, 2009.

E.-J. Rijkhorst, G. Mellema and V. Icke. Blowing up warped disks in 3D. Three-dimensional AMR simulations of point-symmetric nebulae. *Astronomy & Astrophysics*, vol. 444, pp. 849–860, 2005.

E.-J. Rijkhorst, T. Plewa, A. Dubey and G. Mellema. Hybrid characteristics: 3D radiative transfer for parallel adaptive mesh refinement hydrodynamics. *Astronomy & Astrophysics*, vol. 452, pp. 907–920, 2006.

L. F. Rodríguez, P. Carral, S. E. Kurtz, K. Menten, J. Cantó and R. Arceo. Radio Detection of the Exciting Sources of Shell H II Regions in NGC 6334. In J. Arthur and W. J. Henney, editors, *Revista Mexicana de Astronomía y Astrofísica Conference Series*, vol. 15, pages 194–196, 2003.

L. F. Rodríguez, Y. Gómez and D. Tafoya. Changes in the Radio Appearance of MWC 349A. *The Astrophysical Journal*, vol. 663, pp. 1083–1091, 2007.

S. Rosswog and D. Price. *Meshfree Methods for Partial Differential Equations IV*, vol. 65 of *Lecture Notes in Computational Science and Engineering*, chapter 3D Meshfree Magnetohydrodynamics, pages 247–275. Springer-Verlag, 2008.

A. Rudolph, W. J. Welch, P. Palmer and B. Dubrulle. Dynamical collapse of the W51 star-forming region. *The Astrophysical Journal*, vol. 363, pp. 528–546, 1990.

G. B. Rybicki and A. P. Lightman. *Radiative processes in astrophysics*. John Wiley & Sons, Inc., 1979.

E. E. Salpeter. The Luminosity Function and Stellar Evolution. *The Astrophysical Journal*, vol. 121, pp. 161–167, 1955.

M. T. Sandford, II, R. W. Whitaker and R. I. Klein. Radiation-driven implosions in molecular clouds. *The Astrophysical Journal*, vol. 260, pp. 183–201, 1982.

M. Schmidt-Voigt and J. Köppen. Influence of stellar evolution on the evolution of planetary nebulae. I. Numerical method and hydrodynamical structures. *Astronomy & Astrophysics*, vol. 174, pp. 211–222, 1987.

D. Semenov, T. Henning, C. Helling, M. Ilgner and E. Sedlmayr. Rosseland and Planck mean opacities for protoplanetary discs. *Astronomy & Astrophysics*, vol. 410, pp. 611–621, 2003.

P. R. Shapiro and H. Kang. Hydrogen molecules and the radiative cooling of pregalactic shocks. *The Astrophysical Journal*, vol. 318, pp. 32–65, 1987.

F. H. Shu. *The physics of astrophysics*, vol. I: Radiation. University Science Books, 1991.

F. H. Shu. *The physics of astrophysics*, vol. II: Gas dynamics. University Science Books, 1992.

F. H. Shu. Self-similar collapse of isothermal spheres and star formation. *The Astrophysical Journal*, vol. 214, pp. 488–497, 1977.

F. H. Shu, F. C. Adams and S. Lizano. Star formation in molecular clouds: Observation and theory. *Annual Reviews of Astronomy & Astrophysics*, vol. 25, pp. 23–81, 1987.

P. M. Solomon, D. B. Sanders and A. R. Rivolo. The Massachusetts Stony Brook galactic plane CO survey: Disk and spiral arm molecular cloud populations. *The Astrophysical Journal*, vol. 292, pp. L19–L24, 1985.

L. Spitzer, Jr. *Physical Processes in the Interstellar Medium*. John Wiley & Sons, Inc., 1978.

V. Springel. E pur si muove: Galiliean-invariant cosmological hydrodynamical simulations on a moving mesh. *arxiv:0901.4107v2*, 2009.

S. W. Stahler and F. Palla. *The Formation of Stars*. Wiley-VCH Verlag, 2004.

S. W. Stahler, F. Palla and P. T. P. Ho. The Formation of Massive Stars. In V. Mannings, A. P. Boss and S. S. Russell, editors, *Protostars and Planets IV*, pages 327–351. The University of Arizona Press, 2000.

J. M. Stone, E. C. Ostriker and C. F. Gammie. Dissipation in Compressible Magnetohydrodynamic Turbulence. *The Astrophysical Journal*, vol. 508, pp. L99–L102, 1998.

G. Strang. On the Construction and Comparison of Difference Schemes. *SIAM Journal on Numerical Analysis*, vol. 5(3), pp. 506–517, 1968.

B. Strömgren. The Physical State of Interstellar Hydrogen. *The Astrophysical Journal*, vol. 89, pp. 526–547, 1939.

G. Tenorio-Tagle. The Gas Dynamics of H II Regions. I. The Champagne Model. *Astronomy & Astrophysics*, vol. 71, pp. 59–65, 1979.

J. E. Tohline. Hydrodynamic Collapse. *Fundamentals of Cosmic Physics*, vol. 8, pp. 1–81, 1982.

E. F. Toro. *Riemann Solvers and Numerical Methods for Fluid Dynamics*. Springer-Verlag, 1997.

J. K. Truelove, R. I. Klein, C. F. McKee, J. H. Hollman II, L. H. Howell and J. A. Greenough. The Jeans Condition: A New Constraint on Spatial Resolution in Simulations of Isothermal Self-gravitational Hydrodynamics. *The Astrophysical Journal*, vol. 489, pp. L179–L183, 1997.

N. J. Turner and J. M. Stone. A Module for Radiation Hydrodynamic Calculations with ZEUS-2D Using Flux-limited Diffusion. *The Astrophysical Journal*, vol. 135, pp. 95–107, 2001.

N. J. Turner, O. M. Blaes, A. Socrates, M. C. Begelman and S. W Davis. The Effects of Photon Bubble Instability in Radiation-dominated Accretion Disks. *The Astrophysical Journal*, vol. 624, pp. 267–288, 2005.

N. J. Turner, E. Quataert and H. W. Yorke. Photon Bubbles in the Circumstellar Envelopes of Young Massive Stars. *The Astrophysical Journal*, vol. 662, pp. 1052–1058, 2007.

R. K. Ulrich. An infall model for the T Tauri phenomenon. *The Astrophysical Journal*, vol. 210, pp. 377–391, 1976.

D. Van Buren and M.-M. Mac Low. Bow shock models for the velocity structure of ultracompact H II regions. *The Astrophysical Journal*, vol. 394, pp. 534–538, 1992.

D. Van Buren, M.-M. Mac Low, D. O. S. Wood and E. Churchwell. Cometary compact H II regions are stellar-wind bow shocks. *The Astrophysical Journal*, vol. 353, pp. 570–578, 1990.

E. T. Vishniac. The dynamic and gravitational instabilities of spherical shocks. *The Astrophysical Journal*, vol. 274, pp. 152–167, 1983.

M. Walmsley. Dense Cores in Molecular Clouds. In S. Lizano and J. M. Torrelles, editors, *Circumstellar Disks, Outflows and Star Formation*, number 1 in Revista Mexicana de Astronomía y Astrofísica Conference Series, pages 137–148, 1995.

C. Weidner, P. Kroupa and I. Bonnell. The relation between the most-massive star and its parental star cluster mass. *arxiv:0909.1555v1*, 2009.

D. J. Whalen and M. L. Norman. Three-Dimensional Dynamical Instabilities in Galactic Ionization Fronts. *The Astrophysical Journal*, vol. 672, pp. 287–297, 2008.

D. J. Whalen and M. L. Norman. Ionization Front Instabilities in Primordial H II Regions. *The Astrophysical Journal*, vol. 673, pp. 664–675, 2008.

S. C. Whitehouse and M. R. Bate. Smoothed particle hydrodynamics with radiative transfer in the flux-limited diffusion approximation. *Monthly Notices of the Royal Astronomical Society*, vol. 353, pp. 1078–1094, 2004.

S. C. Whitehouse, M. R. Bate and J. J. Monaghan. A faster algorithm for moothed particle hydrodynamics with radiative transfer in the flux-limited diffusion approximation. *Monthly Notices of the Royal Astronomical Society*, vol. 364, pp. 1367–1377, 2005.

A. Whitworth. The erosion and dispersal of massive molecular clouds by young stars. *Monthly Notices of the Royal Astronomical Society*, vol. 186, pp. 59–67, 1979.

A. P. Whitworth and H. Zinnecker. The formation of free-floating brown dwarves and planetary-mass objects by photo-erosion of prestellar cores. *Astronomy & Astrophysics*, vol. 427, pp. 299–306, 2004.

A. P. Whitworth, A. S. Bhattal, S. J. Chapman, M. J. Disney and J. A. Turner. The preferential formation of high-mass stars in shocked interstellar gas layers. *Monthly Notices of the Royal Astronomical Society*, vol. 268, pp. 291–298, 1994.

R. J. R. Williams. Shadowing instabilities of ionization fronts. *Monthly Notices of the Royal Astronomical Society*, vol. 310, pp. 789–796, 1999.

R. J. R. Williams, D. Ward-Thompson and A. P. Whitworth. Hydrodynamics of photoionized columns in the Eagle Nebula, M 16. *Monthly Notices of the Royal Astronomical Society*, vol. 327, pp. 788–798, 2001.

M. G. Wolfire and J. P. Cassinelli. Conditions for the formation of massive stars. *The Astrophysical Journal*, vol. 319, pp. 850–867, 1987.

D. O. S. Wood and E. Churchwell. The morphologies and physical properties of ultracompact H II regions. *The Astrophysical Journal Supplement Series*, vol. 69, pp. 831–895, 1989.

P. Woodward and P. Colella. The Numerical Simulation of Two-Dimensional Fluid Flow with Strong Shocks. *Journal of Computational Physics*, vol. 54, pp. 115–173, 1984.

T. Xie, L. G. Mundy, S. N. Vogel and P. Hofner. On Turbulent Pressure Confinement of Ultracompact H II Regions. *The Astrophysical Journal*, vol. 473, pp. L131–L134, 1996.

H. W. Yorke. The dynamical evolution of H II regions—Recent theoretical developments. *Annual Reviews of Astronomy & Astrophysics*, vol. 24, pp. 49–87, 1986.

H. W. Yorke and P. Bodenheimer. The Formation of Protostellar Disks. III. The Influence of Gravitationally Induced Angular Momentum Transport on Disk Structure and Appearance. *The Astrophysical Journal*, vol. 525, pp. 330–342, 1999.

H. W. Yorke and P. Bodenheimer. Theoretical Developments in Understanding Massive Star Formation. In H. Beuther, H. Linz and T. Henning, editors, *Massive Star Formation: Observations Confront Theory*, vol. 387 of *Astronomical Society of the Pacific Conference Series*, pages 189–199, 2008.

H. W. Yorke and E. Krügel. The Dynamical Evolution of Massive Protostellar Clouds. *Astronomy & Astrophysics*, vol. 54, pp. 183–194, 1977.

H. W. Yorke and C. Sonnhalter. On the Formation of Massive Stars. *The Astrophysical Journal*, vol. 569, pp. 846–862, 2002.

M. Zahn. On the gravitational instability of an infinite homogeneous medium. *American Journal of Physics*, vol. 44, pp. 29–31, 1976.

Q. Zhang, P. T. P. Ho and N. Ohashi. Dynamical Collapse in W51 Massive Cores: CS (3–2) and CH_3CN Observations. *The Astrophysical Journal*, vol. 494, pp. 636–656, 1998.

Q.-F. Zhu, J. H. Lacy, D. T. Jaffe, M. J. Richter and T. K. Greathouse. [Ne II] Observations of Gas Motions in Compact and Ultracompact H II Regions. *The Astrophysical Journal Supplement Series*, vol. 177, pp. 584–612, 2008.

H. Zinnecker and H. W. Yorke. Toward Understanding Massive Star Formation. *Annual Reviews of Astronomy & Astrophysics*, vol. 45, pp. 481–563, 2007.

Danksagung

> Some call it faith, some call it love.
> Some call it guidance from above.
> You are the reason we found ours,
> So thank you, stars.
>
> *(Katie Melua, Thank You, Stars; Lyrics: Mike Batt)*

Ich danke...

- ...meinem Betreuer Ralf Klessen für die ausgezeichnete Unterstützung während meiner Doktorarbeit...

- ...Mordecai Mac Low für die vielen hilfreichen „Telecons" und einen wunderbaren Gastaufenthalt am American Museum of Natural History in New York...

- ...Robi Banerjee für seine ausdauernde Hilfe bei der Benutzung des FLASH-Codes und seiner IDL-Skripte...

- ...Eric Keto, Roberto Galván-Madrid und Henrik Beuther für zahlreiche Diskussionen über Beobachtungen...

- ...Cornelis Dullemond für seine Zusammenarbeit an RADMC-3D...

- ...der ganzen Arbeitsgruppe für die angenehme Arbeitsatmosphäre...

- ...meinen Eltern für den Rückhalt in all den Jahren...

- ...Anna-Sophia für die Ermutigungen in schwierigen Zeiten und das mir entgegengebrachte Vertrauen und Verständnis.

Vielen Dank!

Die VDM Verlagsservicegesellschaft sucht für wissenschaftliche Verlage abgeschlossene und herausragende

Dissertationen, Habilitationen, Diplomarbeiten, Master Theses, Magisterarbeiten usw.

für die kostenlose Publikation als Fachbuch.

Sie verfügen über eine Arbeit, die hohen inhaltlichen und formalen Ansprüchen genügt, und haben Interesse an einer honorarvergüteten Publikation?

Dann senden Sie bitte erste Informationen über sich und Ihre Arbeit per Email an *info@vdm-vsg.de*.

Sie erhalten kurzfristig unser Feedback!

VDM Verlagsservicegesellschaft mbH
Dudweiler Landstr. 99
D - 66123 Saarbrücken
www.vdm-vsg.de

Telefon +49 681 3720 174
Fax +49 681 3720 1749

Die VDM Verlagsservicegesellschaft mbH vertritt

Printed by Books on Demand GmbH, Norderstedt / Germany